C.I.L.R.I.S.

Gears and Gear Manufacture

Join Us on the Internet

WWW: http://www.thomson.com
EMAIL: findit@kiosk.thomson.com

thomson.com is the on-line portal for the products, services and resources available from International Thomson Publishing (ITP). This Internet kiosk gives users immediate access to more than 34 ITP publishers and over 20,000 products. Through *thomson.com* Internet users can search catalogs, examine subject-specific resource centers and subscribe to electronic discussion lists. You can purchase ITP products from your local bookseller, or directly through *thomson.com.*

Visit Chapman & Hall's Internet Resource Center for information on our new publications, links to useful sites on the World Wide Web and an opportunity to join our e-mail mailing list. Point your browser to: **http://www.chaphall.com/chaphall.html** or **http://www.chaphall.com/chaphall/mecheng.html** for Mechanical Engineering

A service of **I(T)P**®

Gears and Gear Manufacture

The Fundamentals

Richard H. Ewert

CHAPMAN & HALL

ITP® INTERNATIONAL THOMSON PUBLISHING

New York • Albany • Bonn • Boston • Cincinnati • Detroit
London • Madrid • Melbourne • Mexico City • Pacific Grove
Paris • San Francisco • Singapore • Tokyo • Toronto • Washington

Cover design: Curtis Tow Graphics

Printed in the United States of America

Chapman & Hall
115 Fifth Avenue
New York, NY 10003

Chapman & Hall
2-6 Boundary Row
London SE1 8HN
England

Thomas Nelson Australia
102 Dodds Street
South Melbourne, 3205
Victoria, Australia

Chapman & Hall GmbH
Postfach 100 263
D-69442 Weinheim
Germany

International Thomson Editores
Campos Eliseos 385, Piso 7
Col. Polanco
11560 Mexico D.F
Mexico

International Thomson Publishing–Japan
Hirakawacho-cho Kyowa Building, 3F
1-2-1 Hirakawacho-cho
Chiyoda-ku, 102 Tokyo
Japan

International Thomson Publishing Asia
221 Henderson Road #05-10
Henderson Building
Singapore 0315

1 2 3 4 5 6 7 8 9 10 XXX 01 00 99 98 97

Library of Congress Cataloging-in-Publication Data

Ewert, Richard H.
Gears and gear manufacture : the fundamentals / Richard H. Ewert.
p. cm.
Includes bibliographical references and index.
ISBN 0-412-10611-6
1. Gearing. 2. Gearing -- Manufacture. I. Title.
TJ184.E94 1996
621.8′33--dc20 96-18802
CIP

British Library Cataloguing in Publication Data available

To order this or any other Chapman & Hall book, please contact **International Thomson Publishing, 7625 Empire Drive, Florence, KY 41042.** Phone: (606) 525-6600 or 1-800-842-3636. Fax: (606) 525-7778. e-mail: order@chaphall.com.

For a complete listing of Chapman & Hall titles, send your request to **Chapman & Hall, Dept. BC, 115 Fifth Avenue, New York, NY 10003.**

To Darle W. Dudley, Gear Consultant, San Diego, CA, friend and mentor of recent years—his guidance and encouragement have been invaluable;

To the late Earle B. Miller, Professor of Mathematics at Illinois College, who instilled in me his fascination for mathematics,

and

To the late Earle Buckingham, Professor of Mechanical Engineering at MIT, who made engineering come alive for me.

Foreword

Gearing plays a role, usually unseen, in the lives of everyone in the civilized world. Few people know anything about gears and even fewer understand them. Even practicing engineers, except those who are gear specialists, know little except the rudiments about gears.

Dick Ewert, with this book, makes a significant contribution, not only in informing students of mechanical engineering of the fundamentals of gearing, but also in providing a reference source for practicing engineers who do not design or specify gears every day but who, on occasion, have a gear requirement.

To this book Dick brings over fifty years of experience in the gear business, as an engineer, as a marketing director, and as the head of a gear manufacturing company. He is a past president of the American Gear Manufacturers Association, and over many years has been a leader in the promotion of AGMA as the accepted voice and technical standard writer of the industry.

I have had the pleasure of being Dick's business partner for thirty years and can attest, maybe better than anyone, to his great technical knowledge in his chosen field of gearing, to his practical approach to engineering solutions, and to his personal and business integrity.

Edward R. Sewall, President
Sewall Gear Mfg. Co.
St. Paul, Minnesota

Preface

This book is written to provide business people dealing with geared machinery a quick and easy explanation of what is involved. The non-expert needs some basic knowledge of gearing so that he or she can handle his or her responsibilities properly. The material in this book will be very helpful to people making management decisions about purchasing gears, planning the production of a new geared product, or in understanding how to take care of geared machinery already in service.

It is written also for practicing engineers, most of whom are not specialists, to provide them with the basics of gearing. I have known many such engineers whose employers are gear users. They need to have some knowledge of gearing but only as one element of the many things they must know. The material in this book will be helpful to them and others who want to know something, but not everything, about gears.

This book will be helpful also to the relative newcomer to gearing, even if this is the most he or she ever learns. If there is desire to delve more deeply into gear engineering and design, then this book can be a stepping stone to further study.

It must be noted here that this book, by itself, is not intended to tell the reader how to design gears. It does, however, discuss some of the pitfalls, and it tells where to look to find out all about gear design.

Immediately following this preface is "The History of Gears" by Darle W. Dudley, Gear Consultant, San Diego, CA. The writer is indebted to Mr. Dudley for his contributions. In addition to writing

The History of Gears he proofread much of my manuscript, offering suggestions for revisions and additions that were most helpful. Also Edward J. Wellauer, retired Vice President/Engineering, Falk Corporation, Milwaukee, offered a number of suggestions that were very helpful, as did Edward R. (Ned) Sewall, by successor at Sewall Gear. In addition I received some suggestions from a "non-gear man," Gregory P. Struve, P.E., Vice President of Robert Muir Company. He suggested some clarifications that he felt would be helpful to readers who have only a minimal knowledge of gearing.

I am indebted also to the late Professor Earle Buckingham of MIT. I have used a number of his definitions (paraphrased) and excerpts from notes taken in discussions with him throughout.

A word of caution: The study of gearing can become addictive. I've been told that many years ago a prominent gear engineer's wife gave birth to a baby girl. They couldn't agree on a name for her, so HE named her "Involute," the name of the curve used for the profiles of most gear teeth.

Richard H. Ewert
St. Paul, Minnesota
December, 1994

Contents

Appendices

The History of Gears

by Darle W. Dudley, P.E., Dudley Technical Group, Inc., San Diego, CA

Meshing gear teeth have been in use by mankind for well over 3000 years.* As a matter of fact we don't know where or when the first gears were used. Gears came out of the mists of antiquity along with other mechanical devices such as the potter's wheel, the inclined plane, pulleys and levers, and many others.

The earliest gears were probably made out of wood. The Greeks did make small gears for astronomical devices out of metal. The Romans used wooden gears in grist mills for power transmission, as well as iron and bronze gears in devices to control motion where very little power was needed.

All through the Middle Ages gears were used in water mills. The mechanical clock invented in the 1300s used small metal gears. Pendulum clocks that flourished in early times were full of gears. Some of these clocks in churches were quite large.

The first really extensive use of power gears occurred in Holland where thousands of windmills pumped sea water out of the land. These giant windmills used a train of wooden gears to take the power from a fast-running wind rotor at the top of the mill and deliver it to a slow-running Archimedes type of screw pump at the bottom of the mill.

As the Industrial Revolution got underway in the 1700s there was

*Dudley, Darle W. "The Evolution of the Gear Art," published by American Gear Manufacturers Association, 1500 King Street, Alexandria, VA 22314, 1969.

a need for all kinds of gear drives. Rather good gears were made in cast iron using patterns that were quite accurately made with wooden teeth, spokes, and hub. Soon even better gears were made, after the first gear-cutting machines were invented and used to mill gear tooth shapes in cast iron or steel gear blanks. In the nineteenth century people considered that a *normal* gear had cast teeth and a *precision* gear had cut teeth.

As the twentieth century began, the automobile and airplane came into being. Already in the late nineteenth century, steam-powered ships, electric-powered trolley cars, and geared machine tools had come into widespread use. Gear drives came into general use in vehicles, mills, and all manner of mechanical devices. The development of gear engineering went hand-in-hand with the very widespread use of gears.

The technology of gearing required a complex array of machinery to make the needed gears, plus engineering rules and data to deal with things such as tooth contact stresses, tooth bending stresses, lubrication of fast-meshing gear teeth, and metallurgical understanding of the metals (or other materials) used in gearing.

What was needed, of course, was a gear design that was practical to make with machines available and a size and quality good enough for the gear drive to last as long as needed in each application. And there was, of course, the usual complication that the gear drive had to be reasonably low in cost and not too large or too heavy!

All during the twentieth century there has been steadily accelerating progress in the gear art. The art of making gears has changed from hand operating machines making one part at a time to automatic machines, computer controlled, that can achieve production rates and quality levels never thought possible in the early part of the century.

The engineering design art for gears has blossomed to the extent that a whole shelf of technical books is needed by the engineer to understand how to do all the geometric calculations applicable to design and have all the needed data on materials and processes. Then much more information is needed on trade standards plus data to cover all the failure hazards that must be avoided to get the required gear life and reliability.

To summarize what has happened in gear technology, my personal experience, after many years in the gear field, leads me to these conclusions:

Over 90% of man's technical knowledge applicable to gearing has been developed in the time period 1890 to 1990—in the last 100 years.

The last 100 years can be divided into four† periods, with each period having approximately equal overall advancement in the gear art:

1890–1930 The initial period in which gear manufacturing and gear engineering became recognized and established.

1930–1960 Gear manufacturing and gear engineering became broad based. The gear art matured substantially.

1960–1980 An era of considerable advancement in gear technology. The gear art became very sophisticated.

1980–1990 The old established gear art saw drastic changes as innovations in machinery design, computer control, new materials, and new lubricants came into being. The traditional way things used to be done in the gear shop or the gear design office have all tended to change substantially.

The gear art has now progressed to one of immense complexity, amazing capability, and a bewildering tendency to make the machinery and design rules of only thirty years ago obsolete! Gear technology is highly advanced now just like the technologies having to do with space vehicles, FAX machines, plastics, lasers, and many other things.

---

†Dudley, Darle W., "Gear Technology–Past, Present and Future," paper presented at International Conference On Gearing in Zhengzhou, Henan Province, China, 1988.

Introduction: The Gear Industry

Most American and Canadian gear companies are members of American Gear Manufacturers Association (AGMA). They range in size from 15 to 20 employees to over 1000. They range in product lines from industrial open gearing, automotive gearing, aircraft gearing, enclosed gear drives, to name a few.

Many of these companies have very sophisticated inspection equipment and procedures. Some have the bare essentials, which is probably acceptable for AGMA quality up to about Q6. Therefore when ordering gears it is important, particularly for an initial order, for the buyer to clearly specify the quality he must have and make sure in some manner that the supplier he has selected knows how to make and inspect them. Up to about AGMA Q8 probably the quality number is all that is needed. For AGMA Q9 and higher it is suggested that it is wise to also show the tolerances required.

Many manufacturers of unassembled gearing have a few customers who, as their usage increases, are tempted to put in their own gear cutting machines. What they often fail to consider is that in order to justify the purchase of, say, a $100,000 hobbing machine, they must have enough work to operate it as many hours as the gear maker runs it, and usually this is two shifts every working day. In addition the gear maker is operating two or more such machines by one man each shift.

To determine the cost per hour of running such a machine at various levels of production, refer to Chapter 16 in this book. Such an analysis will quickly show that for most gear users it is cheaper to buy than to make. I have known very few captive gear shops that operated enough hours to absorb their share of factory overhead costs.

When ordering standard catalogued enclosed drives the buyer must make sure that his selection procedure is in accordance with AGMA practice regarding end use and proper selection of service factor. His selection of a supplier should in large part be based on the supplier's ability to make recommendations in this regard. For standard reducers availability of spare parts should not be a problem, but it is well to make sure of this, as well as to find out how long the parts are available if a reducer line should be discontinued.

For special engineered units such as steel mill roll drives, high-speed high-horsepower drives, and other nonstandard units, spare parts should be ordered with the units. Also, and this is very important, the manufacturer must have the facilities to conduct adequate tests of the units before shipment, to be witnessed by the buyer's engineers.

Gear units having either pitch line speeds over 25,000 feet per minute or 1000 horsepower or more can be a risk to people close to the units. There have been a few incidents around the world where a gear part has failed, burst through the gear casing, and injured or killed people nearby. There must be adequate safety planning so that dangerous situations with powerful gearing are handled in the same manner that other dangerous high-speed machinery is handled. The operator and the control instrumentation should be housed in a concrete steel reinforced booth a safe distance away from the machinery. During initial testing of a new unit, sandbag emplacements may be advisable to protect those getting data to qualify the unit as satisfactory for its intended use.

It should be obvious that similar precautions should be taken when such units are placed into operation. The cost of a breakdown can be enormous, not to mention the tragedy of people getting hurt or perhaps killed.

Regular maintenance inspections must be scheduled and monitored to make sure they are performed.

Gears and Gear Manufacture

1

Gear Types

First, what is a gear?

A *gear* is a machine element, the purpose of which is to transmit motion and power from its shaft, through another gear to its shaft, by means of gear teeth. A usual requirement is that the motion transmitted is uniform. That is, if the driving gear is rotating at a uniform angular velocity, the driven gear must rotate uniformly also. The smaller of two gears in mesh, usually the driver, is called the *pinion*; the larger is the *gear*.

The various types of gears in common use can be categorized as follows:

Parallel Shafts

- Spur
- Helical
- Double–helical
- Herringbone
- Internal
- Rack

Intersecting Shafts

- Straight bevel
- Spiral bevel
- Zerol®

Nonintersecting Nonparallel Shafts

- Hypoid
- Worm gearing

Crossed-axis helical
Spiroid®

Special Gear Types
Elliptical
Triangular
Square
Multiple-sector
Scroll

A description of each of these gear types follows, with illustrations of most of them.

Parallel Shafts

Spur Gears

These are gears with the teeth cut parallel to their axes of rotation. They are the simplest to design and make as compared to other gear types, and so probably most gear designers start with spur gearing for a given application, changing to helicals or herringbones if found to be necessary. Spurs are shown in Figures 1.1 and 1.2.

Figure 1.1. Spur gear and pinion. (Courtesy of Sewall Gear Mfg. Co., St. Paul, MN.)

Figure 1.2. Combination internal–external spur. (Courtesy of Sewall Gear Mfg. Co., St. Paul, MN.)

Helical Gears

Helical gears (pronounced "hellical" by gear people) are gears with the teeth cut at an angle to their axes of rotation. The teeth are sections of a helix that spirals around the body of the gears, not straight as are spur gear teeth.

An important consideration in designing helical gears is to work from standard spur gear pitches. The advantages are twofold: The helix angle can be selected to provide adequate tooth overlap (a rule-of-thumb is 1.2 or greater pitches overlap), and the designer is not limiting his sources of supply. With this approach the pitch of the spur hob becomes the normal pitch of the helical gear. The circular pitch in the transverse direction (direction of rotation) is thus greater than the circular pitch of the hob. It equals the normal pitch divided by the cosine of the helix angle. Normal pitch helical gears are therefore larger in diameter than spur gears of the same pitch and number of teeth.

An alternative is to design to available *transverse pitch* hobs or cutters. This approach results in helical gears with the same numbers of teeth and pitch diameters they would have if designed as spur gears. Therefore transverse pitch helicals can be used to replace spurs when ratio and center distance must not be changed, if provision is made to handle the lateral thrust that occurs with helical gearing.

The helix angles most commonly used for transverse pitch helicals are 7½° and 15°, sometimes 23° and 30°, and 45° for crossed-axis helicals. As indicated, the disadvantages of transverse pitch helicals are that optimum helix angle for tooth overlap cannot be selected and sources of supply are limited to those who have these special hobs. In addition, the tooth strength is less since the normal tooth thickness of transversal pitch helical gearing is less than the normal tooth thickness of helicals designed from spur hobs.

Helical gearing is illustrated in Figures 1.3 and 1.4.

Double Helical Gears

These are related to helicals in that they are in effect two helicals side by side with the hands of the helices opposed to each other. The advantages over single helicals are primarily in drives requiring very few teeth, such as found in steel mill roll drives and in geared pumps, and in drives where axial thrust cannot be tolerated.

The gear blanks for double-helical gears are machined to have a relief groove in the center of the tooth face, between the two helices. Without the relief groove you have continuous tooth *herringbone gears*.

Figure 1.3. Helical gear and pinion. (Courtesy of Sewall Gear Mfg. Co., St. Paul, MN.)

Figure 1.4. Internal gear and pinion (helical). (Courtesy of Sewall Gear Mfg. Co., St. Paul, MN.)

These are usually considered to be less satisfactory than double-helicals for high-speed drives because the apex of the teeth can form an oil trap, causing noise, and at the higher speeds increase heat generation. Also relief grooves minimize tooth breakage caused by axial runout, as continuous teeth will have stress concentration at the apex of the teeth. With little or no axial runout stress concentration is less and continuous teeth are less apt to break than teeth with a relief groove.

The most common helix angle for double-helical and herringbone gearing is 30°, and the pitch designation is in the transverse direction.

The drives for most oil field pumping units are double-helicals or herringbones. One reason for this is that one of the oldest and largest manufacturers of such equipment invested heavily in herringbone gear generating machines years ago. The gear reducers they designed, and build, therefore do not require thrust bearings as would single-helical reducers. Another reason is that the thrust bearings that would have to be used with single-helicals would fail prematurely because of the reversing load direction, which in an oil field pumping unit drive occurs twice per revolution of the low-speed gear. The load reversals loosen the thrust bearings, resulting in dynamic loads on them which they are not designed to handle.

A herringbone gear oil well pumping unit drive is illustrated in Figure 1.5 and an oil field pumping unit in Figure 1.6.

Figure 1.5. Herringbone gear oil well pumping unit drive. (Courtesy of Lufkin Industries, Inc. Lufkin, TX.)

Figure 1.6. Oil field pumping unit. (Courtesy of Lufkin Industries, Inc., Lufkin, TX.)

Internal Gears

The teeth of internal gears are generated from the basic rack, as are the teeth of external spur and helical gears, but their application has certain limitations.

The major uses of internal gears include simple pinion and gear drives, planetary drives, the outer ring of flexible couplings, and internal differential drives, to name a few.

In flexible couplings the internal and external gears have the same number of teeth, so the problems of rotating pinion driving a gear do not exist. In the latter the typical problems of involute interference, tip interference and fillet interference become more acute than with external gears. To avoid these the Buckingham design procedure that follows is recommended.

Internal gears are produced by broaching (up to about 6-inch PD) or shaping (to about 100-inch PD). Over 100-inch PD internals are usually cut with form cutters on machines equipped with a gooseneck attachment that holds the cutter. Most internals for flexible couplings are made in production quantities and are broached.

Rack

Most racks are spur, but can also be helical or double-helical. A rack is straight (infinite pitch radius), used in drives in which rotary to reciprocating motion is required, as on a lathe bed. Such a rack, with the driving pinion, is illustrated in Figure 1.7.

The rack has other uses. As the base circle of a spur gear increases in size it approaches infinity; the involute curve approaches a straight line and eventually becomes a straight sided rack tooth. A rack tooth will, therefore, roll with any gear tooth of the same pitch and pressure angle (the teeth are conjugate with each other), and can therefore be used to generate gear teeth.

Gear cutting tools for spur, helical, and double-helical gears have teeth that are of the *basic rack* form.

Intersecting Shafts

Straight Tooth Bevel Gears

These are gears with teeth cut on the outside of a cone and which, if extended, would intersect at the cone's apex. Since the diameter of a

INTERNAL GEAR DRIVES

THE 20° INVOLUTE FULL DEPTH FORM APPEARS TO BE THE MOST EFFECTIVE STANDARD TOOTH FORM FOR THESE DRIVES. IN ORDER TO AVOID INVOLUTE INTERFERENCE, THE INTERNAL RADIUS OF THE INTERNAL GEAR MUST BE INCREASED OVER THE CONVENTIONAL SIZE FOR THE SMALLER GEARS. IN ORDER TO AVOID INTERFERENCE BETWEEN THE CORNER OF THE INTERNAL GEAR TOOTH AND THE FILLET OF THE PINION TOOTH, ADDITIONAL CLEARANCE MUST BE PROVIDED AT THE TIP OF THE GEAR TOOTH. IN ORDER TO MAINTAIN EFFECTIVE CONTACT, THE SIZE OF THE PINION MUST BE INCREASED OVER THE CONVENTIONAL SIZE. IN ORDER TO GENERATE FULL INVOLUTE PROFILES ON THE INTERNAL GEAR TEETH, THE CUTTER MUST HAVE NOT LESS THAN 16 TEETH.

THE FOLLOWING TOOTH PROPORTIONS, FOR PINIONS OF 16 TEETH AND LARGER, WILL MEET THE FOREGOING CONDITIONS:

WHEN,

R_o = OUTSIDE RADIUS OF PINION
R_1 = PITCH RADIUS OF PINION
N_1 = NUMBER OF TEETH IN PINION
N_2 = NUMBER OF TEETH IN INTERNAL GEAR
R_2 = PITCH RADIUS OF INTERNAL GEAR
R_i = INTERNAL RADIUS OF INTERNAL GEAR
C = CENTER DISTANCE
P = DIAMETRAL PITCH

$$R_1 = \frac{N_1}{2P} \qquad R_2 = \frac{N_2}{2P} \qquad C = \frac{N_2 - N_1}{2P} = R_2 - R_1$$

$$R_o = \frac{N_1 + 2.500}{2P} = R_1 + \frac{1.250}{P} \qquad R_i = \frac{N_2 - 1.200}{2P} = R_2 - \frac{0.600}{P}$$

T_1 = ARC TOOTH THICKNESS OF PINION AT R_1
T_2 = ARC TOOTH THICKNESS OF INTERNAL GEAR AT R_2

$$T_1 = \frac{1.7528}{P} \qquad T_2 = \frac{1.3888}{P}$$

WHEN THE NUMBER OF TEETH IN THE PINION IS LESS THAN 16, THE PINION AND INTERNAL GEAR MUST BE ENLARGED STILL MORE. THE FOLLOWING TABLES GIVE THE PROPORTIONS OF 1 DP PINIONS AND INTERNAL GEARS, PRODUCED BY CUTTERS WITH 16 OR MORE TEETH. FOR OTHER DIAMETRAL PITCHES, THE TABULATED VALUES MUST BE DIVIDED BY THE DIAMETRAL PITCH USED. NO ALLOWANCE FOR BACKLASH HAS BEEN MADE IN THESE TABLES.

TO AVOID TRIMMING WHEN FEEDING THE CUTTER TO DEPTH ON THE INTERNAL GEAR, THE GEAR MUST BE 12 OR MORE TEETH LARGER THAN THE CUTTER.

The above was reproduced, with permission, from *Manual of Gear Design, Section Two, Spur and Internal Gears*, Earle Buckingham and Eliot K. Buckingham, Revised, Buckingham Associates, Springfield, VT, 1980 (Original edition copyright, 1935).

cone is greatest at its base the teeth will be thicker at the base and this is called the *heel* of the teeth. The thinner end of the teeth is called the *toe*. The *cone distance* of a bevel is the length from the base to the apex of the cone. The face of a bevel is the length from the heel to the toe, and should ***NEVER*** be greater than one third the cone

Figure 1.7. Gear rack and pinion (spur). (Courtesy of Sewall Gear Mfg. Co., St. Paul, MN.)

distance. The pitch of a bevel is measured at the heel. The shaft angle of bevel gears is usually 90°, although bevel gearing can be designed and cut to operate on any shaft angle. A straight-tooth bevel gearset is illustrated in Figure 1.8.

Two bevel gears with equal numbers of teeth designed to run with each other are called *miter gears*.

Figure 1.8. Straight bevel gears. (Courtesy of The Gleason Works, Rochester, NY.)

Most straight-tooth bevel gears today are generated with crowned teeth. These are called *Coniflex*® bevels, a Gleason Works trademark. The bevels in Figure 1.8 are Coniflex bevels. The initial tooth-bearing pattern of Coniflex teeth will be an elongated "bubble" shape, which should, when assembled, be located nearer the toe than the heel of the teeth.

There are many machines still in operation that cannot cut Coniflex teeth. Bevels cut on these machines should be cut to have an initial toe bearing and assembled so as not to lose it, as the toe end of the teeth will deflect under load more than the heel, and will have better running-in characteristics.

Bevel Gears—Angular

These are bevels whose axes are not at right angles.

Bevel Gears—Crown

A crown gear is a bevel with a plane pitch surface. The crown gear is analogous to the basic rack in spur gears.

Bevel Gears—Formate

These are bevels in which the gear member has nongenerated teeth, usually with straight-sided tooth profiles. The pinion member has generated teeth that are conjugate to the mating gear teeth.

Bevel Gears—Revacycle

These are gears cut on the Gleason Revacycle® machine. The teeth are conjugate to each other but are not interchangeable with gears cut by other methods. Their most common use is in automobile bevel gear differentials.

Bevel Gears—Skew

These are bevels with straight teeth, cut at an angle to an element of the pitch cone.

Bevel Gears—Spiral

These are bevels in which the teeth are curved and generated at an angle to an element of the pitch cone, with the angle selected (usually 35°) so the entering tooth comes into contact before the leaving tooth lets go. These are the "helical" version of bevel gears, specified for applications that require more quiet operation than can be achieved with straight tooth bevels.

A spiral bevel gearset is illustrated in Figure 1.9.

Bevel Gears—Zerol®

These are spiral bevels with 0° spiral angle. A possible advantage over Gleason Coniflex straight bevels is that each tooth has one convex and one concave surface, while a Coniflex tooth has two convex surfaces. The initial tooth bearing area of Zerol bevel gearing can therefore be greater than with Coniflex bevels.

It should be noted that both spiral and Zerol bevels can be specified

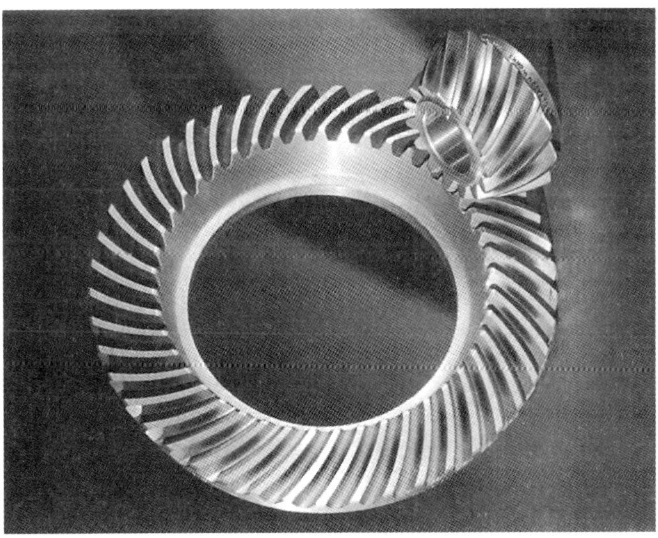

Figure 1.9. Spiral bevel gears. (Courtesy of The Gleason Works, Rochester, NY.)

AGMA Q10 or better as it is possible to grind the teeth on these types, on Gleason bevel gear grinding machines. Straight-tooth bevels cannot be ground.

Nonintersecting Nonparallel Shafts

Hypoid Gearing

These are spiral bevels with the pinion shaft offset, so the shafts do not intersect. Their most common use is in automobile rear axle drives. A hypoid gearset is shown in Figure 1.10.

Worm Gearing

This is a gearset of which the driving member (the worm) is of screw thread configuration. The axis of the worm is usually at 90° to the worm gear shaft.

Figure 1.10. Hypoid gears. (Courtesy of The Gleason Works, Rochester, NY.)

Properly designed and produced, worm gearing provides a good trouble-free drive for smooth operation and permits a wide selection of ratios, but there are precautions that must be taken for optimum gear life. For example, with a high *lead angle* (say 30°), the gear should be tangentially hobbed as infeeding the hob will remove some of the active profile. Also, it is desirable to specify higher pressure angles for the higher lead angles—25 to 30° PA for lead angles over 30° or so. Selection of materials is critical. SAE 65 bronze with about 1% nickel is good for the gear, chill cast, or centrifugally cast if the configuration of the gear permits. The worm should be hardened steel with the threads ground.

Care in assembly is critical, too. The worm gear should be adjusted laterally so the tooth bearing pattern is offset toward the "leaving" side of the teeth, as determined from the direction of rotation of the worm.

Finally, for optimum results, worm gearing should be run in after assembly, at no load to start, gradually increasing to full load or until a polished area appears on the gear teeth. This may not be necessary if the initial tooth bearing area is satisfactory, but if it is less than about 50% of the face width there could be problems without a run-in.

The capacity of a worm gear drive can be increased by designing it at *recess-action* worm gearing. The improved characteristics result from the increase in the recess (polishing) portion of the thread-tooth contact. This is achieved in worm gearing by hobbing the worm gear undersize, not by making the driving worm larger, as one might at first assume from the discussion on spur and helical gear modification. The reason for this is that the worm thread is the basic rack form no matter what its diameter, so increasing the worm diameter will change only the lead angle, not thread-tooth action.

For an illustration of a worm and gear set see Figure 1.11.

Worm Gearing—Double Enveloping

In double-enveloping worm gearing the worm is shaped like an hourglass, so it will be in contact with more worm gear teeth than a conventional worm. This type of gearing will transmit more power than a conventional worm and gear set of the same size, but more heat will be generated in high-speed operation and assembly time will be longer, as worm alignment and position are more critical than with conventional type worm gearing.

Figure 1.11. Worm gearing. (Courtesy of Sewall Gear Mfg. Co., St. Paul, MN.)

For an illustration of a double-enveloping worm gear set see Figure 1.12.

Crossed-Axis Helical Gears

These are helicals with the helix angles of the same hand, usually designed to operate on a 90° shaft angle. These are sometimes called *spiral gears*. They have an initial point tooth contact and hence are limited to relatively light load applications.

An illustration of a crossed-axis helical gearset is shown in Figure 1.13.

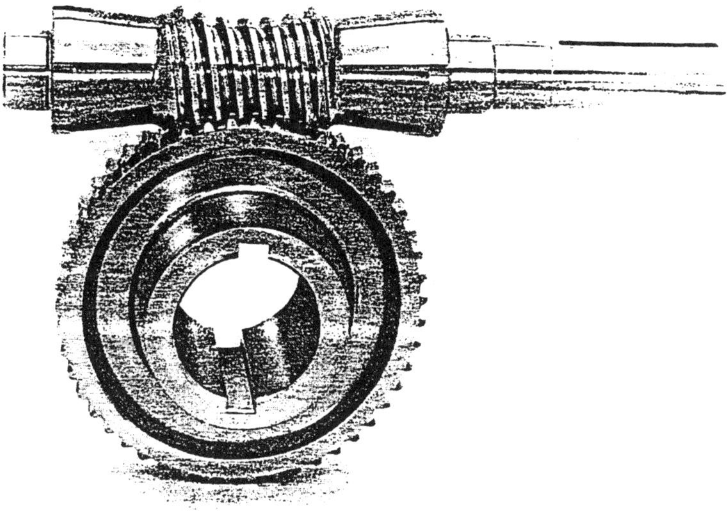

Figure 1.12. Double-enveloping worm gear set. (Courtesy of Ex-Cell-O Corp., Cone Drive Div., Traverse City, MI.)

Figure 1.13. Crossed-axis helicals. (Courtesy of Sewall Gear Mfg. Co., St. Paul, MN.)

Spiroid® Gears

This is a family of gears in which the tooth design is in an intermediate zone between bevel, worm, and face gear design. The Spiroid design is patented by Illinois Tool Works, Chicago.

An illustration of a Spiroid gearset is shown in Figure 1.14.

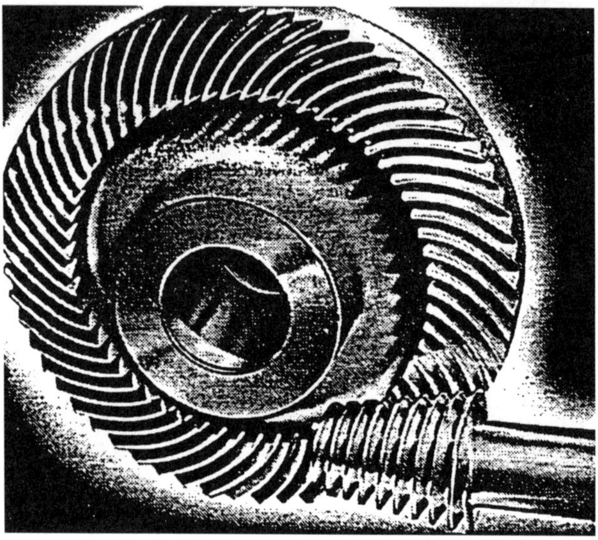

Figure 1.14. Spiroid gearset. (Courtesy of Illinois Tool Works, Spiroid Division, Chicago, IL.)

Special Gear Types

These are used in mechanisms that require a variable or intermittent motion of the output shaft from a constant-speed input. There are many such requirements and many shapes of such gearing are possible. A few of the most common are discussed and illustrated here.

It should be noted that in general such gearing must be operated at relatively low speeds and to transmit motion but at relatively low power.

In small quantities such gearing is usually cut a tooth at a time with range cutters. One of my early experiences in the gear business was

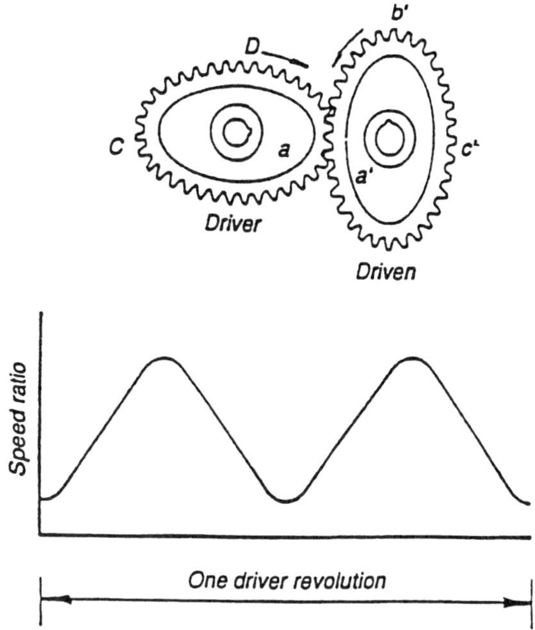

Figure 1.15. Elliptical gears. (Figures 1.15 to 1.19 are reproduced from *Dudley's Gear Handbook*, edited by Dennis P. Townsend, McGraw-Hill, New York, 1991.)

with a gear company in Chicago. A customer called to say he wanted to buy a set of elliptical gears. I didn't even know there was such a thing, but I found out that we had made them before so I took the order. I was curious and wanted to see how we made them, so followed their progress through the shop.

We had an old-timer in the shop who knew how to cut elliptical gear teeth. As I recall he used a milling machine, hand indexing the gear being cut, and cutting one tooth space at a time. It took a long time, and after 16 hours or so my old-timer completed the first gear—a 6″ or so major diameter steel elliptical gear, about 1″ face and about 30 teeth. When he finished that first gear he was a nervous wreck—and he still had one to go!

Elliptical, square, and triangular gears can be cut on Fellows gear shapers, using cutters that are the same shape and size as the gear to

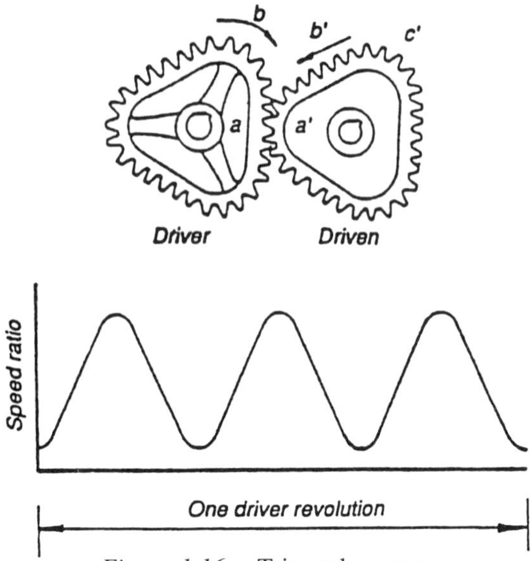

Figure 1.16. Triangular gears.

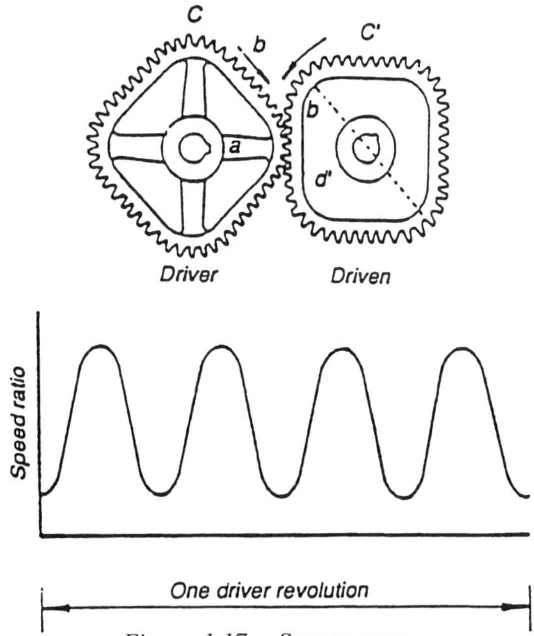

Figure 1.17. Square gears.

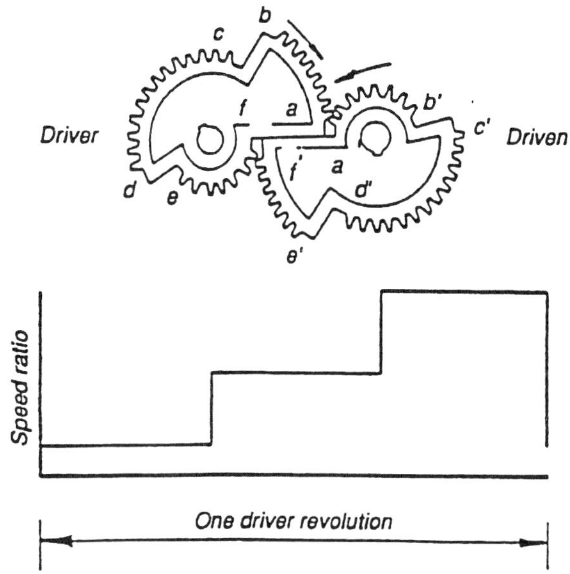

Figure 1.18. Multiple-sector gears.

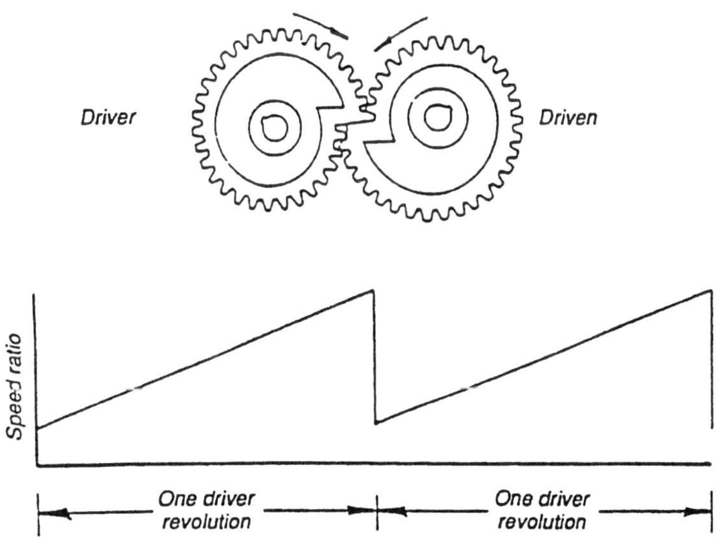

Figure 1.19. Scroll gears.

be cut. Such cutters are expensive, a justified expense only if the quantities to be produced are large enough. Possibly shaper cutters can be designed and made to cut other configurations as well.

Other ways these special gear types can be made are in molded plastic, die cast, or in powdered metal. Each of these requires handmade dies which are very expensive, so here again large quantities are necessary to justify the expense of the dies.

Figures 1.15 to 1.19 show several more or less common examples of these special gear types, with a curve of the type of motion produced with one revolution of the driver gear for each gear type. The illustrations show the following gear types: elliptical, triangular, square, multiple-sector, and scroll.

2

The Involute Curve

The requirement of uniform angular velocity (noted at the beginning of Chapter 1) dictates the use of mating tooth forms that are *conjugate* to each other (from Webster's New World Dictionary, a definition of conjugate is "reciprocally related and interchangeable as to properties, as two points, lines, etc."). Of those that meet this requirement the *involute* curve is the easiest to produce in the shop and is therefore used as the profile for most gear teeth. It is important, then, to have an understanding of some of the properties of the involute curve, the mathematical expression for which is derived on p. 25.

The involute is the curve traced by the end of a taut string being unwound from a nonrotating circle. In gearing this circle is called the *base circle*. The illustrations in Figures 2.1a and 2.1b show elements of a single involute curve and construction of involute curves. The latter shows involutes from two base circles with radii R_{b1} and R_{b2}, respectively and various positions of the string being unwound from each base circle, called *generating lines*. It is seen that the unwinding string is always normal to the involute curve it is generating and is tangent to its base circle at any position it is in.

The involute curves from both base circles in Figure 2.1b are drawn so the involutes are in contact with each other. Note that the generating line from R_{b1}, when in the position *AP*, continues on and is tangent also to R_{b2} at *D*. The intersection of this common tangent with the centerline of the two base circles defines the *pitch point (P)*, and by similar triangles it is seen that the pitch radii (R_1 and R_2) are directly proportional to the base circle radii.

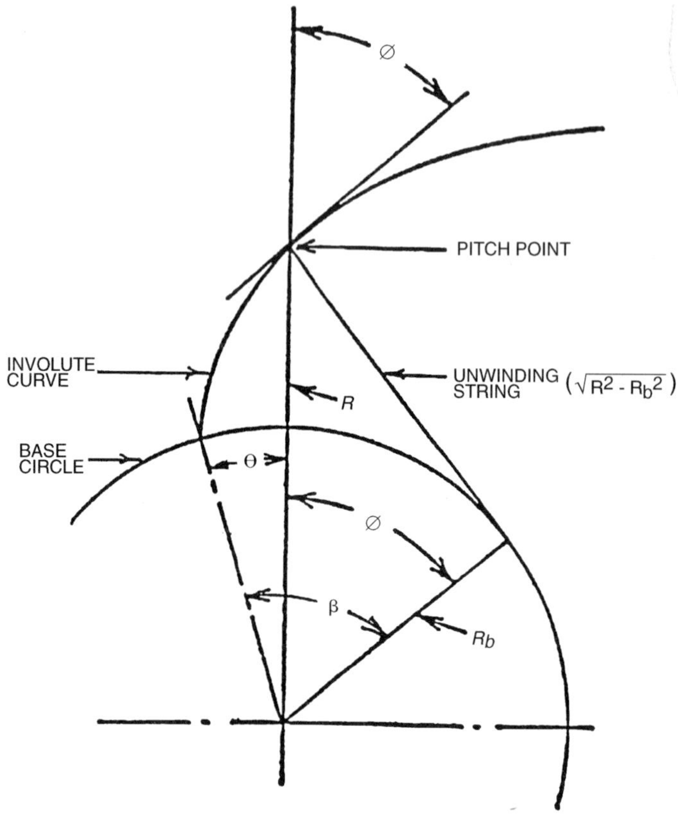

Figure 2.1a. Elements of a single involute curve. R_b = radius of base circle; R = radius to involute curve (pitch radius); θ = angle from start of involute curve to radius r (vectorial angle); ϕ = pressure angle at radius R; β = angle through which string has been unwound.

If you will visualize what happens as these gears rotate, you will see that the point of tooth contact is always along the common tangent to the two base circles (line *AD*), and so this is called the *path of contact* or *line of action*.

The angle ϕ is the *pressure angle* (PA), so called because as you see it is the angle through which the pressure, or power, is transmitted from the driving to the driven teeth.

We speak often of *contact ratio*. This is easy to define, and visualize,

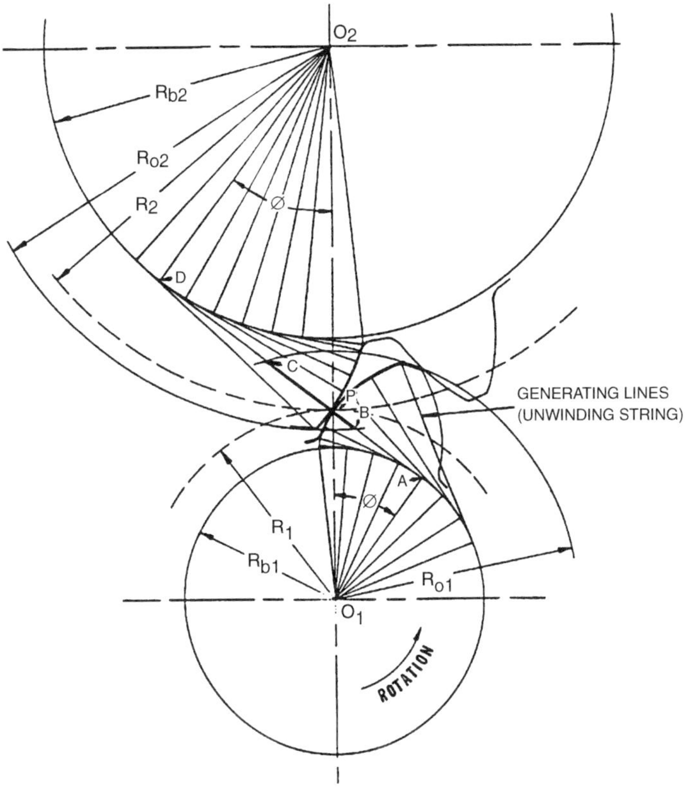

Figure 2.1b. Construction of involute curves.

from Figure 2.1b. If we draw the circles representing the outside radii (R_{o1} and R_{o2}) of these two gears, they intersect the line of action at points *C* and *B*. We see that contact can take place only between these two points. The *arc of action* is the arc through which one tooth travels from the time it first makes contact with its mating tooth until it ceases to be in contact, and is, therefore, the length *BC* measured at the radius of the base circle.

Contact ratio is the length *BC* divided by the *base pitch*. Base pitch is the distance between two successive involutes measured along the circumference of the base circle, so it is the circumference of the base circle divided by the number of teeth. When the contact ratio is less than unity, which it happens to be in the illustration, it means that it is geometrically impossible for more than one tooth to be in contact at a time (for spur gears). For good operating characteristics, the contact ratio of a spur gearset should generally be about 1.4 or greater for relatively high-speed gearing (about 2000 ft per minute *pitch line velocity* or greater), or gearing that must roll as smoothly as possible regardless of speed. A contact ratio of perhaps 1.2 for low-speed gearing might be acceptable but, if possible, it should be greater than this.

By comparing layouts of this type it can be shown that contact ratio increases as pressure angle decreases. So in the pressure angles we usually talk about, 14½° and 20° PA, 14½° PA gears have a greater contact ratio than 20°, for a given number of teeth in pinion and gear, assuming no *undercut* (see Chapter 3). This means that 14½° PA gearing with no undercut would operate more quietly than 20° PA gearing, assuming both are made to exactly the same degree of accuracy.

Note that the radius of curvature of the involute increases as the pressure angle increases. If we compare 14½° and 20° PA gears, it is over 40% greater for 20°, and this greater radius of curvature of the 20° PA tooth enables it to withstand greater surface pressure than the 14½° PA tooth. At the same time the 20° PA tooth is thicker at the *root diameter* and will therefore withstand greater bending loads than the 14½°.

For most applications the 20° PA gear has more advantages than the 14½° and has therefore become more prevalent. In recent years the 25° PA has gained acceptance for heavily loaded relatively low-speed drives.

One might wonder why a pressure angle of 14½° became a standard in the first place. Why not 20°, which became a standard later, or 15°, or maybe 18°? The reason is not very scientific or profound. In the old days coarse pitch gears had cast teeth. Laying out the teeth in the patterns was simpler for 14½° PA gears because the sine of 14½° is .25038, close enough to the fraction ¼ that the pattern maker could use the latter to lay out gear teeth.

Referring to Figure 2.1a, the involute curve is expressed mathematically as follows:

Let

R_b = radius of base circle
R = radius to any point of involute

θ = vectorial angle
β = angle through which string has been unwound.

Then

$$\theta = \beta - \tan^{-1} \frac{\sqrt{R^2 - R_b^2}}{R_b}$$

The length of the generating line $\sqrt{R^2 - R_b^2}$ is the length of the circumference of the base circle which is subtended by the angle β.

Hence

$$\sqrt{R^2 - R_b^2} = R_b\beta, \quad \text{or} \quad \beta = \frac{\sqrt{R^2 - R_b^2}}{R_b}$$

$$\therefore \qquad \theta = \frac{\sqrt{R^2 - R_b^2}}{R_b} - \tan^{-1} \frac{\sqrt{R^2 - R_b^2}}{R_b}$$

This is the polar equation of the involute curve.

Since the first term equals tan ϕ and the second term equals ϕ, this equation can be written:

$$\theta = \tan \phi - \phi = \text{Inv } \phi$$

or

$$\text{Inv } \phi = \tan \phi - \phi$$

If ϕ is the angle in degrees, then:

$$\text{Inv } \phi = \tan \phi - \frac{\pi \phi}{180}$$

From Figure 2.1a we also have the relationship between the pitch and base circles, which is:

$$R_b = R \cos \phi$$

The term "Inv ϕ" is known as the *involute function*. Involute functions are tabulated in most gear books, one of the first of which was *Manual of Gear Design* by Earle Buckingham. The involute function tabulation starting on p. 204 in the appendix is reproduced from that book.

It is interesting to note that Prof. Buckingham hand lettered these and other tables in his manual. The Professor didn't trust the typesetters of his day, so these tables were reproduced directly from the original manuscript by the photo offset process, ensuring accurate reproduction of these and other tables.

Involute functions are used in various formulas relating to involute gear teeth. Three of these, which relate to tooth thickness, are shown in Chapter 5, pp. 49 and 50.

3

Gear Tooth Action

The remarkable, versatile geometric curve defined in Chapter 2, the *involute* curve, can do a lot of things. Its properties provide the means to calculate gear tooth dimensions, gear tooth modification limitations, and measurement and inspection of gear teeth, to name a few.

In addition, it is the basis in the design and manufacture of gear tooth cutting tools, as covered in Chapter 4.

As mentioned previously, its unique properties make it possible to be reproduced as gear tooth profiles in the shop much easier than other curves that have been used, one of which is the *cycloidal* form.

First we shall talk about *modification* and *undercut*. For pinions with small numbers of teeth, the generating tool will cut away that portion of the involute that is near and below the base circle (R_{b1} in illustration on Figure 2.1b).

This cutaway portion is called *undercut* and it increases as the number of teeth decreases. For 14½° PA gears with standard *full depth* teeth, undercut occurs at about 32 (and fewer) teeth. For 20° PA full-depth spur gears it occurs at about 18 (and fewer) teeth, and for 25° PA spur gears it occurs at about 12 (and fewer) teeth.

Undercut pinions will therefore have a less-than-standard active tooth profile, so contact ratio is reduced. Also, undercut reduces tooth strength as it decreases tooth thickness at the root of the teeth. It is, therefore, an important consideration in gear design. It is minimized, or sometimes eliminated altogether, by hobbing or shaping small pinions on oversize diameters. A limiting factor in the amount of modification possible is the diameter at which the tooth would come to a point.

The pinion shown in Fig. 2.1b is a modified pinion. This is evident from the relatively long *addendum* shown ($R_{01} - R_1$), as compared to the *working depth* below the pitch line ($R_{02} - R_2$). On a nonmodified pinion, addendum and working depth below pitch line are equal. The distance from the pitch line to the bottom of the tooth space (the *root*) is the *dedendum*.

The formula to be used when it is necessary to determine the arc tooth thickness at a radius other than the radius at which the tooth thickness is known is:

When (see Figure 3.1):

r_1 = given radius
ϕ_1 = pressure angle at r_1
T_1 = arc tooth thickness at r_1
r_o = outside radius (standard)
r_3 = modified outside radius
r_2 = radius where tooth thickness is to be determined
ϕ_2 = pressure angle at r_2
T_2 = arc tooth thickness at r_2.

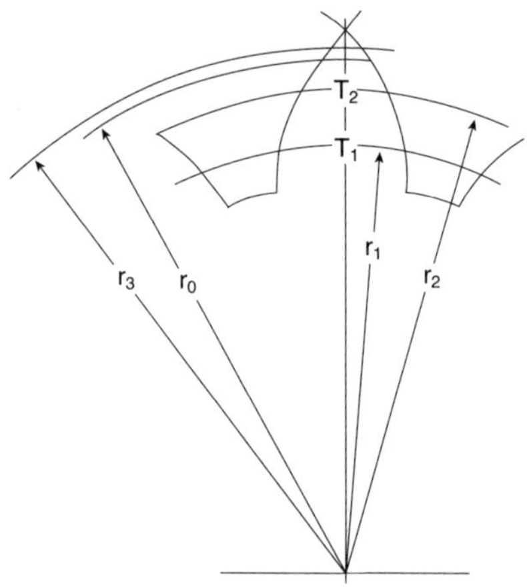

Figure 3.1.

Then:

$$\cos \phi_2 = \frac{r_1 \cos \phi_1}{r_2}$$

$$T_2 = 2r_2 \left[\frac{T_1}{2r_1} + \text{Inv } \phi_1 - \text{Inv } \phi_2\right]$$

This formula is useful in determining width of top land by substituting r_o (outside radius) for r_2. Semi-recess action modification (see Figure 3.3) would be achieved with the outside radius (r_3) about midway between r_2 (see p. 31) and the standard outside radius.

Thus:

$$r_3 = \frac{r_2 - r_3}{2} + r_o$$

To determine the diameter at which the pinion tooth would come to a point, when cut on an oversize diameter, the following formula will do the trick.

When the arc tooth thickness and pressure angle at a given radius are known, to find the radius at which the tooth becomes pointed, proceed as follows (see Figure 3.2):

When
r_1 = given radius
r_2 = radius at which tooth becomes pointed
T_1 = arc tooth thickness at r_1
ϕ_1 = pressure angle at r_1
ϕ_2 = pressure angle at r_2.

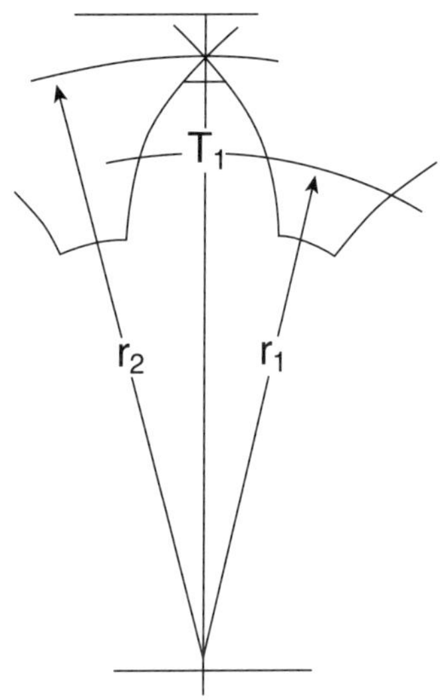

Figure 3.2.

Then

$$\text{Inv } \phi_2 = \frac{T_1}{2r_1} + \text{Inv } \phi_1$$

$$r_2 = \frac{r_1 \cos \phi_1}{\cos \phi_2}$$

Example: Consider a 30 *T*, 6 *DP* (.5236″ *CP*), 14½° PA gear.

$$r_1 = 2.500'' \quad T_1 = .2618 \quad \phi_1 = 14.500°$$

$$\text{Inv } \phi_1 = \tan \phi_1 - \frac{\pi\phi_1}{180} = .25862 - .25307 = .00555$$

$$\text{Inv } \phi_2 = \frac{0.2618}{2 \times 2.500} + .00555 = .05791$$

$$\phi_2 = 30.693° \quad \cos \phi_2 = .85991 \quad \cos \phi_1 = .96815$$

$$r_2 = \frac{2.500 \times .96815}{.85991} = 2.8147$$

For a nonmodified pinion and gear the full-depth addendum is 1/*DP*. The dedendum is 1.157/*DP* in the Brown & Sharpe system, 1.25/*DP* in the AGMA full-depth system, 1.35/*DP* for full-depth teeth to be shaved or ground, and 1.40/*DP* in the full-depth fillet root system for teeth to be induction hardened by scanning. For stub tooth addenda and dedenda see p. 32.

Oversize modification of pinions is desirable for another reason also. It increases the *arc of recess* (length *PC* on p. 23) and reduces the *arc of approach* (length *BP*) when the pinion is the driver. The tooth action through the arc of approach tends to scuff the tooth surfaces under heavy loads and the action through the arc of recess is a polishing action (provided the drive is properly lubricated). So properly lubricated recess action gearing will operate more smoothly and have longer life than gearing with standard tooth proportions. See Figure 3.3 for illustrations of standard vs. recess action gearing.

Note, however, that if the gear drives the pinion, as it would in a speed increaser, and if the pinion is oversize, the intended recess action would become approach action. Another case of gear driving the pinion is in a hoist drive when the hoist is letting the load descend. In general gearing in such drives should have teeth of standard proportions, or if the pinion must be modified oversize to increase tooth strength, then the gear must be modified oversize also, to make the angles of approach and recess as nearly equal as possible.

When a gearset of standard proportions is to be replaced by a modified gearset usually the standard center–center distance must be held. In such a case the gear is cut on an undersize diameter—undersize by the same amount the pinion diameter has been increased. When this is done we have *long-and-short addendum* gears.

A special case when modification of the driving pinion is almost always advisable is for nonmetallic pinions—nylon, fabric-impregnated bakelite, and so forth. Here tooth deflection under load is greater than with metal pinions, and modification minimizes the tendency for the

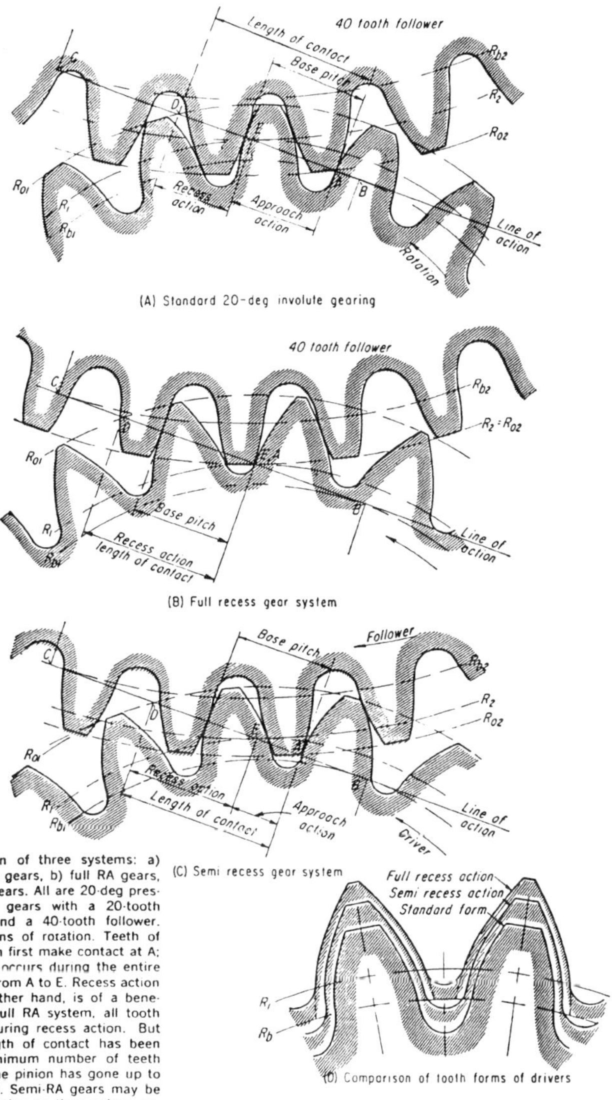

. . Comparison of three systems: a) standard involute gears, b) full RA gears, and c) semi-RA gears. All are 20-deg pressure angle spur gears with a 20-tooth pinion (driver) and a 40-tooth follower. Note the directions of rotation. Teeth of the standard form first make contact at A; detrimental wear occurs during the entire approach action from A to E. Recess action E to D, on the other hand, is of a beneficial nature. In full RA system, all tooth contact occurs during recess action. But because the length of contact has been reduced, the minimum number of teeth permissible for the pinion has gone up to 20 teeth or more. Semi-RA gears may be the best compromise as they reduce approach action yet increase contact length.

Figure 3.3. Standard vs. recess action gears.

---

tips of the gear teeth to gouge into the flanks of the pinion teeth as the latter deflect.

Sometimes it is best to avoid modification of small pinions and live with the undercut you get, or specify *stub teeth*, or perhaps even use *form-cut* (also called *rotary-cut*) pinion and gear (see p. 37). An example is open gearing which is lubricated by heavy grease applied with a stick. We have seen that pinion modification increases the arc of recess, and in this case recess action can be a problem as it is a wiping action of the pinion teeth on the gear teeth. It will tend to wipe away the grease lubricant faster than a nonmodified pinion, since on the latter the arcs of recess and approach are equal. So modification is not recommended for drives where the lubricant used is so heavy that it will not flow back quickly. For such drives the stub tooth is sometimes the answer. Undercut is minimized, or sometimes eliminated, because the root diameter is greater, but contact ratio is less because a stub tooth has a shorter active profile. So unless the number of teeth in the gear is quite large stub teeth can be troublesome. Properly designed stub tooth gearing is good, however, for relatively low-speed heavily loaded drives.

There are two stub tooth systems, both with 20° PA. The first was the *Fellows Stub*, designated as a fraction, such as 4/5 *DP*, which says that the pitch of the gear is 4 DP with a 5 *DP* tooth depth. A later system is the *AGMA 20° Stub* system. In this system the addendum is .8/*DP* and the dedendum is 1/*DP*. The AGMA system is interchangeable with the Fellows system and has largely replaced it.

4

The Cutting of Gear Teeth

In this chapter we give you more examples of the versatility of the involute curve: its use in the design and manufacture of gear tooth cutting tools.

The first example is in cutting gear teeth by *generating*. As the base circle gets larger and larger, its radius approaches infinity, and the involute curve approaches a straight line and eventually becomes a straight-sided *basic rack* tooth. A rack tooth will, therefore, roll with any gear tooth of the same pitch and pressure angle (the teeth are conjugate to each other). If, therefore, the cutting tool is made to the basic rack form, it is capable of generating gear teeth in any diameter gear.

One such cutting tool is the *hob*, a tool of screw thread configuration with flutes milled normal to the thread, or threads, to provide the cutting edges. Its cutting action is that of continuous milling, the gear blank rotating while the hob, also rotating, is moved across the face of the gear, or in hobbing worm gears moved radially or tangentially. A single thread hob cutting a 50-tooth gear would therefore have to rotate 50 times faster than the gear it is cutting.

A typical hob for cutting spur or helical gears and a worm gear hob are illustrated in Figures 4.1 and 4.2.

A Fellows shaper cutter is another type of tool. Its configuration is that of a gear, usually about 4 to 6 inches diameter, with the outside diameter (OD) and teeth tapered to provide clearance for the cutting edges (see Figure 4.3). Its tooth form is the involute curve corresponding to the number of teeth and pressure angle of the cutter. Its cutting

Figure 4.1. Spur gear (or helical) hob. (Courtesy of Sewall Gear Mfg. Co., St. Paul, MN.)

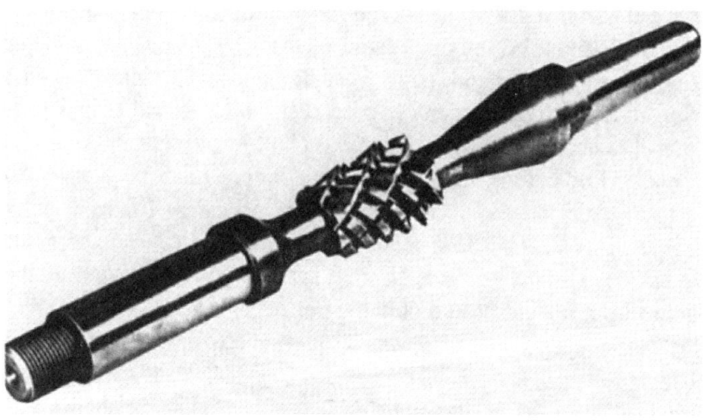

Figure 4.2. Worm gear hob-tapered (for tangential feed), seven starts. This hob was used for hobbing the worm gear shown in Figure 1.11. (Courtesy of Sewall Gear Mfg. Co., St. Paul, MN.)

action is that of shaping as it reciprocates across the gear face, with both the gear it is cutting and the cutter rotating, in the speed ratio of the number of teeth in the gear being cut divided by the number of teeth in the cutter.

Spur, helical, and double-helical gears can be shaped. Cutters for shaping helicals must have the same helix angle as the gear being cut.

Figure 4.3. Shaper cutter. (Courtesy of Sewall Gear Mfg. Co., St. Paul, MN.)

The cutters for shaping external helicals must be the opposite hand of the gear being cut, and for cutting internal helicals the same hand as the gear being cut. For internal gears the number of teeth in the cutter must not exceed that which would trim a portion of the involute, which becomes critical when the number of teeth in the internal is relatively small. For cutting helicals the gear shaper must be fitted with *guides* to provide rotary motion to the cutter as it reciprocates. Internal gears MUST be shaped, as well as gears whose teeth abut a shoulder or another gear. When the latter it is called a *cluster gear*. In such cases a relief groove must be provided in the gear blank for cutter and chip clearance.

Maag gear cutting machines cut teeth with a shaping action also, using cutters that are of basic rack configuration.

A common finishing process is *shaving*, used after hobbing or shaping to improve finish and accuracy. Gears to be shaved must be cut with *preshave* hobs or shaping cutters, which remove stock below the *active profile* of the teeth to provide clearance for the tips of the shaving cutter's teeth. Shaving cutters are essentially helical gears with serrated teeth to provide the cutting edges. The cutter drives the gear being shaved as it is moved back and forth axially while rotating. A shaving cutter is illustrated in Figure 4.4.

Another finishing process is grinding the profiles of the teeth, used to remove heat treat distortion on hardened gears. Grinding is required on hardened gears specified as AGMA Quality 10 or better. Honing or lapping gear teeth can be used to finish and remove minute profile

Figure 4.4. Shaving cutter. (Courtesy of Sewall Gear Mfg. Co., St. Paul, MN.)

errors on hardened gears. A honing machine works like a shaving machine, with the hone a nonmetallic gear impregnated with abrasive particles.

Lapping of spur gear teeth by running two gears together with an abrasive lapping compound is a "last resort" method to remove hardening distortion, except for large gears and when performed by an expert. Excessive lapping of spurs will destroy the involute profile, as when spur gear teeth are in mesh, rolling action takes place only at the pitch point and sliding action above and below the pitch point, so most of the stock removal occurs above and below the pitch point.

A straight-tooth bevel gear generating machine uses two single shaper cutters or, in faster machines, two rotary interlocking cutters, with the cutting edges constrained to move in the path that the teeth of a mating gear would follow. Such cutters are illustrated in Figures 4.5, 4.6 and 4.7.

A spiral bevel gear generating machine uses a rotating face-milling cutter, fed into the workplace, with cutter and work rotating together to provide the generating motion required. Such a cutter is illustrated in Figure 4.8. Gleason spiral bevel machines also cut Zerol® bevels (spiral bevels with 0° spiral angle) and hypoid gears.

Crowned teeth are often specified for spur, helical, and straight-tooth bevel gears. They are thicker at or near the center of the tooth face than at their ends. Crowning is accomplished by crown hobbing, crown shaving, or crown grinding. On straight-tooth bevels the trade

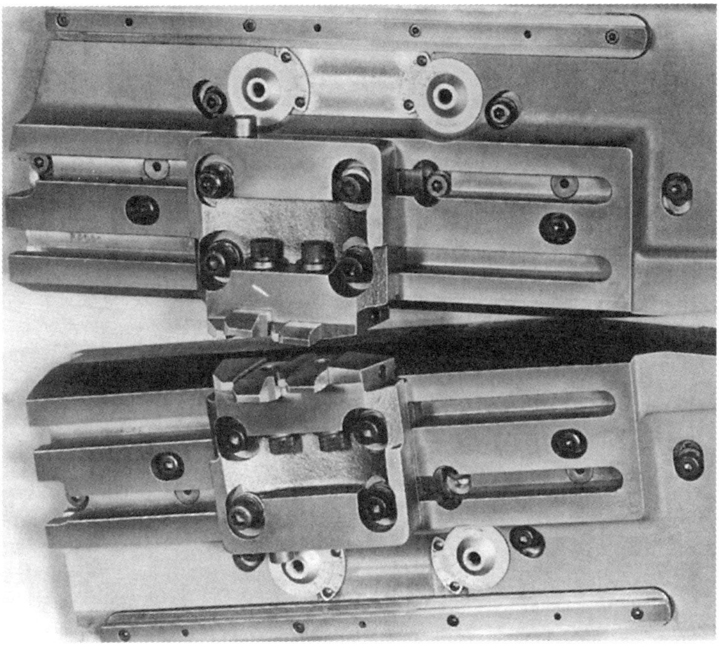

Figure 4.5. Gleason reciprocating cutters, for use with Gleason reciprocating tool generators Nos. (14) 429 and (24) 434, for cutting straight-tooth Coniflex® bevel gears. (Courtesy of The Gleason Works, Rochester, NY.)

designation of crowned teeth is Coniflex®, a Gleason Works trademark. The purpose of crowning is to minimize the chance of end bearing of the teeth when there is misalignment in assembly, distortion from heat treating or deflection under load.

The amount of crown is commonly in the neighborhood of 0.0003 inches per inch of face (0.0006 inch change in tooth thickness per inch of face). It can be varied but if assembly conditions call for more than this, the designer must be concerned with the compressibility of the tooth surfaces and tooth deflection. If the crown exceeds the amount the teeth will compress and deflect under full load to provide adequate tooth bearing area, the full face of the gearing cannot be used in determining the load rating.

Something should be said about *form cutting* (sometimes called *rotary cutting*) of spur gears. Form cutters are milling cutters that have

Figure 4.6. Gleason interlocking cutters, for use with Gleason interlocking tool generators Nos. (104) 439 and (114) 442, for cutting straight-tooth Coniflex® bevels.

the form of a tooth space (see Figure 4.9). They produce an approximate involute form. In general, gears produced by this method are not too satisfactory except for low pitch line velocity operation and in relatively heavy pitches, and in those cases of open gearing, referred to previously, for which modification should be avoided, and strength requirements dictate that undercut cannot be tolerated.

Stamdard form cutters are *range cutters*, so called because there are eight cutters in a set, or fifteen in a set for greater profile accuracy, each of which is designed to cut a range of numbers of teeth. Cutters for pinions in the undercut range are modified so they can mill below the base circle radius. Gear tooth cutters are modified to produce gear teeth that will clear this lower portion of the pinion tooth profile. Since the profiles of form cut teeth are made up of more than one curve they are called *composite* tooth forms. The involute portion of the composite

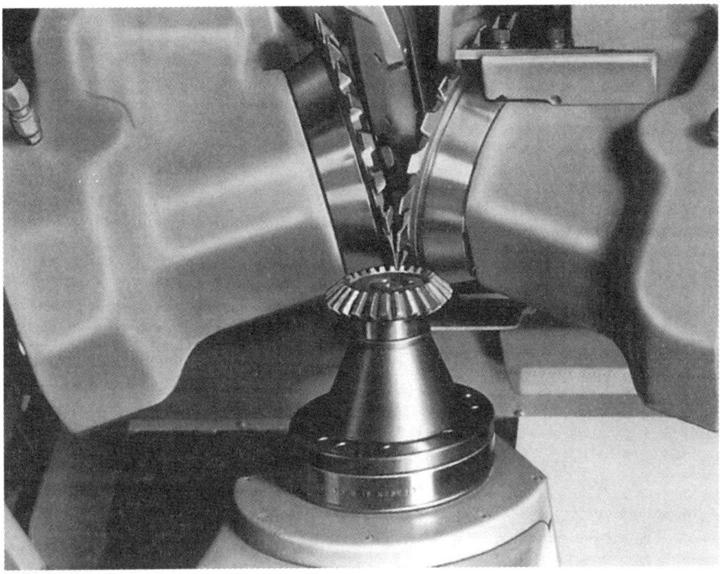

Figure 4.7. Interlocking cutters in operation. (Courtesy of The Gleason Works, Rochester, NY.)

form is correct for the minimum number of teeth in each range. Range cutters are listed in Table 4.1.

In general, form cut gearing is not interchangeable with generated tooth gearing, except in sets.

Gear rack in pitches heavier than about 3 *DP* is usually form cut, because most rack generating machines are not designed to cut the heavier pitches. It is best to use single-purpose rack cutters to cut true involute teeth (straight sides on rack teeth), with some *tip relief* to minimize the possibility of tip interference. A rack and pinion are pictured in Figure 1.7.

Some manufacturers of medium to heavy pitch rack use No. 1 range cutters (135 teeth to a rack). As previously noted this produces a tooth profile that a 135-tooth gear would have, so it is not correct for a rack. The practice of making all racks with *basic rack form* teeth assures smoother rolling with generated tooth pinions.

A finer pitch rack (finer than about 3 *DP*) can be form cut but usually is generated. Rack shapers are used for generating gear rack teeth,

Figure 4.8. Gleason face-mill cutter, for cutting spiral, Zerol® bevel and hypoid gears. (Courtesy of The Gleason Works, Rochester, NY.)

Figure 4.9. Form cutter. (Courtesy of Sewall Gear Mfg. Co., St. Paul, MN.)

Table 4.1. Range Cutters for Cutting Spur Gears

Cutter number	Range of number of teeth, 8 cutter set	Range of number of teeth, 15 cutter set
1	135 to rack	135 to rack
1½	—	80 to 134
2	55 to 134	55 to 79
2½	—	42 to 54
3	35 to 54	35 to 41
3½	—	30 to 34
4	27 to 34	26 to 29
4½	—	23 to 25
5	21 to 26	21 and 22
5½	—	19 and 20
6	17 to 20	17 and 18
6½	—	15 and 16
7	14 to 16	14
7½	—	13
8	12 and 13	12

using the same type of reciprocating cutters used for cutting spur or helical gears by shaping.

An interesting application of a heavy pitch gear rack is in the manufacture of large-diameter gears, larger than can be produced on conventional hobbing machines. One application for such gears is their use as the final drive to transmit low-speed, high-torque motion to the parabolic shaped reflectors of large radio telescopes and satellite tracking antennas. Such structures with reflectors of from 30 to 150 ft diameters and larger have been built. These final drive gears must be large in diameter to provide a high reduction ratio, and to keep gear tooth loads resulting from inertia, momentum, and wind forces within practical limits. The largest antenna drive gear known to the writer has 1980 teeth and is 110 ft in pitch diameter, a full circle gear which is the azimuth drive for a large antenna. Figures 4.10 and 4.11 show a typical tracking antenna and a close-up of the final drive gear, which in this case is the elevation drive and is a 120° sector of about 30 ft pitch radius. Figure 4.12 shows the three-antenna complex of which the antenna in Figure 4.10 is the one on the left. The azimuth drive gear for this antenna is a gear about 72 inches in diameter, cut on the OD of a bearing. The reflector diameter of this antenna is about 60 ft.

Figure 4.10. A typical tracking antenna. (Courtesy of Harris Corporation, Melbourne, FL and Sewall Gear Manufacturing Company, St. Paul, MN.)

These gears are made in segments cut as straight gear racks, then mechanically formed to the required radius. The cutting tools used are form cutters, with the pitch, pressure angle, and profile nonstandard, so that after the rack is formed to its specified radius its pitch becomes standard and its tooth profile becomes a true involute, conjugate to

Figure 4.11. Antenna final drive gear. (Courtesy of Harris Corporation, Melbourne, FL and Sewall Gear Manufacturing Company, St. Paul, MN.)

Figure 4.12. Three-Antenna Complex. The antenna shown in Figure 4.10 is one of the three in the background of this picture, taken near Los Cruces, New Mexico in 1979, built by Harris Corp. for Western Union. The guy in the foreground is your author. I was there to attend the dedication of this antenna complex. At the dedication dinner and ceremony the MC, a Harris vice president, introduced each of about fifty subcontractors in attendance. He called on me first, explaining that "If the gears hadn't worked, none of us would be here."

conventionally hobbed pinion teeth. Assembled with the aid of jigs to set the tooth spacing at the joints between the segments, the gearing operates with remarkable angular uniformity and with tooth bearing patterns typical of generated tooth gearing.

Other gear tooth producing methods are rolling, broaching, die casting, and stamping. Such methods are used primarily for relatively fine pitch spur gearing, in quantities large enough to justify the high cost of the tooling required. Such methods are not suitable for gearing that must be highly accurate.

5

Gear Measurement and Inspection

Gears must be inspected during their manufacture and when they are completed.

Inspection procedures and methods will vary with the size, type, and the specified accuracy of the gear. These are described in the following pages.

Inspection Procedures & Methods

It is common practice to check spur gears by taking measurements of diameters over pins, which are not pins at all, but are short sections of hardened and ground steel rods of various diameters. To measure a gear a pin is laid in tooth spaces opposite each other and diameter measurements are made over the tops of the pins. If the gear has an even number of teeth the pins are placed 180° apart; if an odd number the pins are placed as close as possible to 180° apart. The diameter of the pin to be used is selected to suit the pitch of the gear to be checked. It must be of such size so that it will rest in the tooth space, touching the tooth flanks at or slightly above the pitch line, stick out above the OD of the gear, and must not touch the root diameter.

Formulas for calculating pin measurements are given on pp. 47 and 48. Standard pin, or roll, sizes have been established, and gear makers and users can buy standard size pins for the commonly used pitches of teeth. There are published tables to enable determination of measurements over pins for the standard pitches without going through lengthy calculations.

Table 5.1 gives the standard pin sizes which are commonly used.

Helical gears also can be checked over pins, although theoretically balls should be used instead of pins because a helical tooth is curved. The calculations should be made for ball measurement, but most gear manufacturers then use pins for the actual measurement. There is a slight difference between the calculated ball measurement and the actual pin measurement, the difference increasing as the helix angle increases, but is usually so slight that it can be ignored.

For heavy pitch gears, gear tooth vernier calipers are often used, instead of pins, to measure tooth thicknesses. They measure chordal tooth thickness, from which circular tooth thickness can be calculated if desired for determination of backlash. See pp. 49 and 50 for calculation methods for tooth thickness.

For large-diameter gears it is necessary also to measure the chordal length over a number of teeth, from which tooth thickness and accuracy of tooth spacing can be determined. For such measurements a *span measuring tool* is used, set to touch the flanks of teeth at the ends of each span at or near the middle of the tooth height. A span measuring tool will normally be set by a vernier and will be equipped with a dial indicator to indicate any deviation from the theoretical chord length.

Table 5.1. Pin Sizes Used to Check the Tooth Thickness and Diameters of Spur Gears[a]

Type of tooth	Pressure angle	Pin diameter constant
External, standard or near standard proportions	14½° to 25°	1.727 1.920 1.680
External, long-addendum pinion	14½° to 25°	1.920
Internal, standard proportions	14½° to 25°	1.680 1.440

[a]The pin diameters change as the tooth size changes. For metric design multiply the pin diameter constant above by the module. For English system design, divide the pin diameter constant by the diametral pitch.

Pitch Size Categories

The pitch size categories of "fine," "medium," and "heavy" are usually considered to be as follows:

Category	*DP*	CP (DP/3.1416)	TT (nominal) (*CP*/2)
Fine	20 *DP* and finer	.1571″ and finer	.0785″ and finer
Medium	19 *DP* to 4 *DP*	.1653″ to .7854″	.0826″ to .3927″
Heavy	3½ *DP* to 1 *DP*	.8976″ to 3.1416″	.4488″ to 1.5708″

Calculation of Pin Measurements

Given the arc tooth thickness and pressure angle of a spur at a given radius, to determine the measurement over pins (Figure 5.1), when the number of teeth is even:

T_1 = arc tooth thickness at r_1
r_1 = given radius
ϕ_1 = pressure angle at r_1
ϕ_2 = pressure angle at center of pin
M_1 = measurement over pins
x = radius of pin
r_2 = radius to center of pin
N = number of teeth in gear

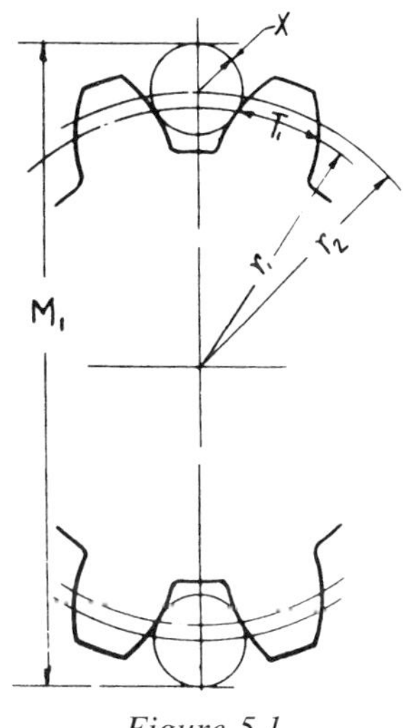

Figure 5.1.

Then,

$$\text{Inv } \phi_2 = \frac{T_1}{2r_1} + \text{Inv } \phi_1 + \frac{x}{r_1 \cos \phi_1} - \frac{\pi}{N}$$

$$M_1 = \frac{2r_1 \cos \phi_1}{\cos \phi_2} + 2x$$

When P = diameter pitch, used pin diameter 1.68/P or slightly larger.

Example: Consider a 30 T, 6 DP, 14½° PA gear.

$$r_1 = 2.500'', T_1 = 0.2618'', \phi_1 = 14.500°, x = .1406'', N = 30$$

$$\text{Inv } \phi_2 = \frac{.2618}{2 \times 2.500} + .00554 + \frac{0.1406}{2\text{–}500 \times 0.96815} - \frac{\pi}{30} = 0.01127$$

$$\phi_2 = 18.272° \quad \cos \phi_2 = .94958$$

$$M_1 = \frac{2 \times 2.500 \times .96815}{0.94958} + 2 \times 0.1406 = 5.37898''$$

When the number of teeth is odd (Figure 5.2):

The formula for Inv ϕ_2 is the same as for even number of teeth.

$$r_2 = \frac{r_1 \cos \phi_1}{\cos \phi_2} \quad M_2 = 2\left(r_2 \cos \frac{90°}{N} + x\right)$$

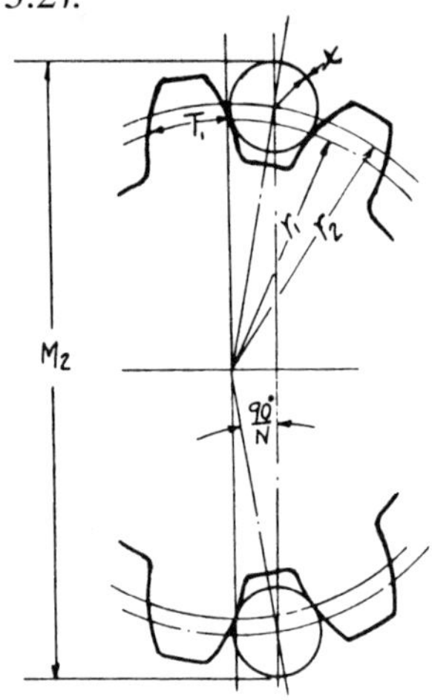

Figure 5.2.

Example: Consider a 31 T, 6 DP, 14½° PA gear.

$$r_1 = 2.5833'', T_1 = .2618'', \phi_1 = 14.500°, x = .125'', N = 31$$

$$\text{Inc } \phi_2 = \frac{.2618}{2 \times 2.5833} + .00554 + \frac{.1406}{2.5833 \times .96815} - \frac{\pi}{31} = .01109$$

$$\phi_2 = 18.178°$$

$$\cos \phi_2 = .95009 \quad \cos \frac{90°}{N} = \cos 2.90323° = 0.99872$$

$$r_2 = \frac{2.5833 \times .96815}{.95009} = 2.6324$$

$$M_2 = 2(2.6324 \times .99872 + .1406) = 5.53926''$$

Calculations for Tooth Thickness

Gear tooth verniers measure the chordal tooth thickness, so this must be calculated from the known circular, or arc, tooth thickness. The formula to do this is:

When

T = arc tooth thickness at r
T_c = chordal tooth thickness at r

Then

$$\text{arc } \beta = \frac{T}{2r} \text{ radians}$$

$$T_c = 2r \sin \beta$$

See Figure 5.3.

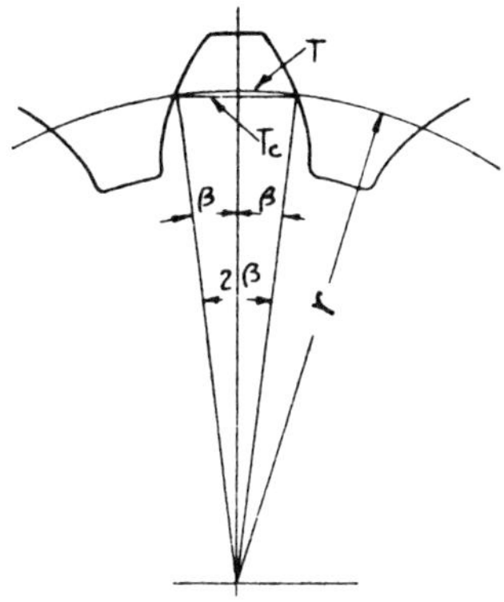

Figure 5.3.

Example: Consider a 30 T, 6 DP (.5236″ CP), 14½° PA gear.

$$r = 2.500'' \quad T = .2618''$$

$$\text{arc } \beta = \frac{.2618}{5.000} = .05236 \text{ radians} = 3.000°$$

$$\sin \beta = .05234$$

$$T_c = 2 \times 2.500 \times 0.05234 = 0.2617''$$

Calculations for Chordal Addendum

It is then necessary also to find the height of the chordal addendum. The formula to do this is:

When

T = arc tooth thickness at r
T_c = chordal tooth thickness at r
r = given radius
r_o = outside radius
a_c = chordal addendum.

Then

$$\text{arc } \beta = \frac{T}{2r}$$

$$a_c = r_o - \cos \beta$$

See Figure 5.4.

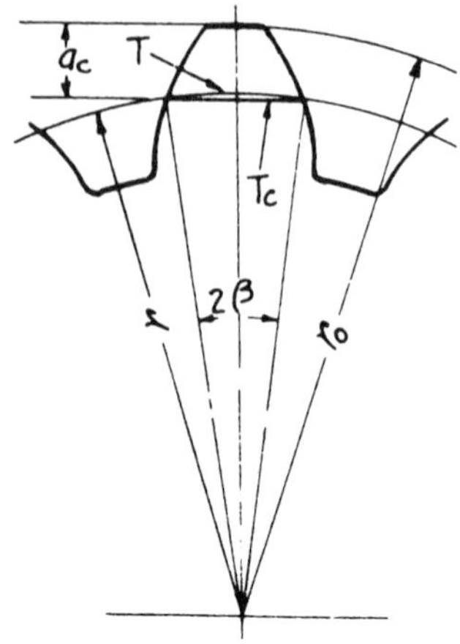

Figure 5.4.

Example: Consider a 30 *T*, 6 *DP* (0.5236″ *CP*), 14½° PA gear.

$$r = 2.500'' \quad r_o = 2.6667'' \quad T = .2618''$$

$$\text{arc } \beta = \frac{.2618}{5.000} = .05236 \text{ radians} = 3.000°$$

$$\cos \beta = 0.99863$$

$$a_c = 2.6667 - 2.500 \times 0.99863 = .1701''$$

Measurement of bevel gearing is accomplished by mounting gear and pinion on a Gleason bevel gear checking machine and rolling them together, driving the gear by the pinion, with a brake on the gear shaft to provide a load. Red lead is applied to the gear teeth, enabling examination of tooth bearing area and position, and measuring of backlash and runout.

Worms are checked over pins. Worm gears are checked by meshing with a mating worm on a fixture and checked for tooth bearing area and position, and for backlash. For repeat or production jobs it is desirable to use a master worm, made to a high degree of accuracy.

In addition to the span measuring tool described on p. 46 and the bevel gear checking machine described above, other gear inspection and testing equipment, and inspection methods, are listed in the following pages.

Involute Checking Machine

These machines utilize the basic "unwinding string" theory of the generation of the involute curve. A probe follows the tooth profile of the gear being checked as it rotates. At the same time the theoretically exact involute curve is generated by unwinding a steel band from a cylinder having the same diameter as the base circle of the gear. The deviation of the actual from the exact is recorded, telling the inspecter whether the involute curve of the gear being checked is within specified tolerances.

Lead Checking Machine

These machines are used to check the accuracy of the helix angle on helical and herringbone gears, the lead angle on worms, and to make sure that there is no helix angle on spur gears.

Optical Comparator

This is a device that projects the image of a gear tooth being checked with the image of a theoretically correct gear tooth, permitting comparison of the two.

Worm Gear Checking Fixture

This is a fixture, usually designed and built by the gear manufacturer or user, to hold a mating worm and gear on their operating center distance for checking tooth bearing area and location of the bearing area, and for checking backlash. For production jobs it is desirable to use a master worm, hardened and ground, made to a high degree of accuracy. Worms are checked by making pin measurements, with the pins laid in the spaces between the worm thread, or threads.

6

Backlash

It is important to understand that all types of gears must have *backlash.* This is the gap between the profiles of meshing teeth when the driving side of a tooth is held tight against the driven at operating center distance. It can be measured by a feeler gauge.

A rule-of-thumb, useful as a guide, for determining the amount of backlash for various pitches, is 0.030 inch to 0.050 inch divided by the diametral pitch. Using this rule a 2 *DP* gearset would have 0.015 inch to 0.025 inch backlash, a 3 *DP* gearset 0.010 inch to 0.017 inch, and so on. In practice, however, the designer must think in terms of tooth thicknesses and tooth thickness (*TT*) tolerances, and then decide what to specify for each gear. In addition, *runout* and center distance tolerance have to be considered, as each will cause the effective tooth thicknesses at the operating pitch circle to vary. Actual backlash, therefore, is what results from these variables after a gearset is assembled. It should not be specified as such until the designer has determined that the backlash he has in mind is consistent with the tolerances he has specified.

To illustrate, the backlash which results from applying data from AGMA Standard 390.03, *AGMA Gear Handbook,* Vol. 1 (see Appendix, pp. 164 to 170), and assuming Quality 8C, 2 *DP* spur gearing, not modified, 12-inch P*D* pinion and 50-inch P*D* gear, can be determined as shown in Table 6.1.

The foregoing comments on backlash and Table 6.1 apply to industrial gearing up to about AGMA Quality 10. In addition there are many applications where heat expansion and cold contraction must be

Table 6.1. Determination of Backlash

	Pinlon (″)	Gear (″)
Runout tolerance (total indicator reading)	.0075″ (max.)	.0105″ (max.)
Corresponding effect on tooth thickness (TT)	+.0027″ −0.0027″	+.0038″ −0.0038″
Specified TT tolerance (assumed)	+.0000″ −0.0048″	+.0000″ −0.0048″
Center distance tolerance (assumed)	+0.0050″ − .0000″	
Corresponding effect on TT[a]	+.0000″ −0.0018″	+.0000″ −0.0018″

The resulting total effective tooth thickness variation, measured at the operating pitch circles, would be as follows:

Runout effect on TT	+.0027″ −.0027″	+.0038″ −.0038″
TT mfg. tolerance	+.0000″ −.0048″	+.0000″ −.0048″
Center distance effect on TT	+.0000″ −.0018″	+.0000″ −.0018″
Variation resulting from above	+.0027″ −.0093″	+.0038″ −.0104″
TT reduction for runout	−.0027″ −.0027″	−.0038″ −.0038″
TT reduction for 0.015″ min. (assumed) B/L	−.0075″ −.0075″	−.0075″ −.0075″
Total TT variation at pitch circles	−.0075″ −.0195″	−.0075″ −.0217″

Gearsets made to these specifications, with center distance tolerance as indicated, would therefore operate with backlash of .015 inch minimum to .041 inch maximum.

[a]Variation in depth of mesh changes effective tooth thickness in the ratio of 1 to .73 for 20° PA.

considered, and relative expansion and contraction between gears and housings when they are of dissimilar materials.

One example is marine gearing—the propeller drive gears that reduce the speed of the prime movers (steam turbines or diesel engines) to the much lower speeds of the propellers. Most navy and many commercial ships are driven by steam turbines, which operate at much higher speeds than diesel engines. On these the first reduction gears get up to speed quickly at start-up, generating heat at the mesh which takes a while to dissipate through the lubricant and then through the housing. This heat will cause the first reduction gearing to expand, often by enough to reduce backlash substantially. Therefore when these gears are manufactured, their teeth must be cut thinner than would be considered normal for industrial gearing.

Most ship propulsion gearing is double-helical, some single-helical, designed with finer pitches than would normally be used for industrial gearing of the same size. For example, a typical marine gear of 100 inches *PD* would probably be about 4 *DP*, where a typical industrial gear would be more like 2 *DP*. The reason for this is that the finer pitch gearing will operate more quietly than the heavier pitch, important to the navy in its efforts to minimize sound that could be picked up by enemy vessels. Gear noise is important to operators of passenger ships as well. They don't want to hear complaints about gear noise from passengers in cabins located above the engine room.

An example where gears and housings are of dissimilar materials is gearing in army ground transportation vehicles that operate in a wide variety of temperatures, from the tropics to the arctic. Most of these gears are enclosed in aluminum or magnesium housings which have different coefficients of expansion than steel. Therefore the expansion coefficients of both gears and housings must be considered when specifying backlash.

Still another example is aerospace gearing, the most highly loaded of which is the gearing driving propellers and helicopter rotors. These are of necessity lightweight, are made from premium steels, carburized and hardened, and teeth finished by grinding. They are usually enclosed in aluminum or magnesium housings. Here again, therefore, the coefficients of expansion of both the gears and the housings must be considered when specifying backlash.

It should be noted that backlash is in no way related to gear quality. One example is the aerospace gearing described herein. It is AGMA Q13 to Q15, but is required to have more backlash than most commercial quality gearing.

7

Gear Materials and Inspection of Materials

Materials

The materials from which gears are made are the steels; the cast irons; the bronzes; aluminum; and the nonmetallic materials such as the phenol resins, nylon, etc. Of these steel is the most widely used. You will see why from the wide variety of types and properties available discussed in this chapter.

We start with our discussion on the steels, the low carbons of which are carburizing grades and the 0.40 to 0.50 carbons are through-treating grades.

Typical examples of such steels are 8620 and 4140. Higher strength steels often specified are 4320 and 4340. The last two digits in the numbers represent the points of carbon present, so 8620 is low carbon and 4140 is medium carbon (a "point" of carbon is .01%, so 4140 steel contains .40% of carbon). The first two digits indicate the type of alloying elements in the steel. Tables 7.1 and 7.2 list typical gear steels, the first giving the common heat treat practice and hardenability characteristics of each, the second the chemical composition of each. If the first two digits are "10" there are no alloying elements (except manganese), and these are called "carbon steels."

As noted above reference is made to two of the common heat treat practices used in the heat treating of steel gears, most of which are subjected to some form of heat treatment. Details of the most common heat treatments used are provided in Chapter 8.

Table 7.1. Typical Gear Materials—Wrought Steel

Common alloy steel grades	Common heat treat practice[a]	General remarks/application
1045	T-H, 1-H, F-H	Low hardenability
4130	T-H	Marginal hardenability
4140	T-H, T-H&N, 1-H, F-H	Fair hardenability
4145	T-H, T-H&N, 1-H, F-H	Medium hardenability
8640	T-H, T-H&N, 1-H, F-H	Medium hardenability
4340	T-H, T-H&N, 1-H, F-H	Good hardenability in heavy sections
Nitralloy 135 Mod.	T-H&N	Special heat treatment
Nitralloy G	T-H&N	Special heat treatment
4150	1-H, F-H, T-H, TH&N	Quench crack sensitive Good hardenability
4142	1-H, F-H, T-H&N	Used when 4140 exhibits Marginal hardenability
4350[b]	T-H, 1-H, F-H	Quench crack sensitive, excellent hardenability in heavy sections
1020	C-H	Very low hardenability
4118	C-H	Fair core hardenability
4620	C-H	Good case hardenability
8620	C-H	Fair core hardenability
4320	C-H	Good core hardenability
8822	C-H	Good core hardenability in heavy sections
3310[b]	C-H	Excellent hardenability (in heavy sections) for all three grades
4820	C-H	
9310	C-H	

[a]C-H = carburize harden; T-H = through-harden; F-H = flame harden; T-H&N = through-harden then nitride; I-H = induction harden.

[b]Recognized, but not current standard grade.

This table and Table 7-2 are reproduced from AGMA 2004-B89, *Gear Materials and Heat Treatment Manual,* with the permission of the publisher, the American Gear Manufacturers Association, 1500 King Street, Suite 201, Alexandria, VA 22314.

Table 7.2. Standard Alloy Steels—Chemical Composition Limits (%)

AISI or SAE Number	C	Mn	P (Max)	S (Max)	Si	Ni	Cr	Mo
1045	0.43/0.50	0.60/0.90	0.040	0.050	—	—	—	—
4130	0.28/0.33	0.40/0.60	0.035	0.040	0.15/0.35	—	0.80/1.10	0.15/0.25
4140	0.38/0.43	0.75/1.00	0.035	0.040	0.15/0.35	—	0.80/1.10	0.15/0.25
4145	0.43/0.48	0.75/1.00	0.035	0.040	0.15/0.35	—	0.80/1.10	0.15/0.25
8640	0.38/0.43	0.75/1.00	0.035	0.040	0.15/0.35	0.40/0.70	0.40/0.60	0.15/0.25
4340	0.38/0.43	0.60/0.80	0.035	0.040	0.15/0.35	1.65/2.00	0.70/0.90	0.20/0.30
4150	0.48/0.53	0.75/1.00	0.035	0.040	0.15/0.35	—	0.80/1.10	0.15/0.25
4142	0.40/0.45	0.75/1.00	0.035	0.040	0.15/0.35	—	0.80/1.10	0.15/0.25
4350	0.48/0.53	0.60/0.80	0.035	0.040	0.15/0.35	1.65/2.00	0.70/0.90	0.20/0.30
1020	0.18/0.23	0.30/0.60	0.040	0.050	—	—	—	—
4118	0.18/0.23	0.70/0.90	0.035	0.040	0.15/0.35	—	0.40/0.60	0.08/0.15
4620	0.17/0.22	0.45/0.65	0.035	0.040	0.15/0.35	1.65/2.00	—	0.20/0.30
8620	0.18/0.23	0.70/0.90	0.035	0.040	0.15/0.35	0.40/0.70	0.40/0.60	0.15/0.25
4320	0.17/0.22	0.45/0.65	0.035	0.040	0.15/0.35	1.65/2.00	0.40/0.60	0.20/0.30
8822	0.20/0.25	0.75/1.00	0.035	0.040	0.15/0.35	0.40/0.70	0.40/0.60	0.30/0.40
3310	0.08/0.13	0.45/0.60	0.025	0.025	0.15/0.30	3.25/3.75	1.40/1.75	—
4820	0.18/0.23	0.50/0.70	0.035	0.040	0.15/0.35	3.25/3.75	—	0.20/0.30
9310	0.08/0.13	0.45/0.65	0.025	0.025	0.15/0.35	3.00/3.50	1.00/1.40	0.08/0.15

The .40 to .50 carbon grades are suitable for flame or induction hardening as well as through-hardening. It should be noted that usually induction hardening is preferable to flame hardening, particularly for tooth-to-tooth hardening of the heavier pitches, because with the induction process closer control of hardness pattern and depth is possible. When hardening of the roots as well as the flanks of the teeth is required, gears to be induction hardened should be cut with hobs that produce full fillet roots. It should be noted also that large steel gears to be induction hardened should be weldments with rolled steel rims, not castings. Castings can, and often do, have inclusions that cause cracks when quenched.

An important factor to consider when specifying steel to be used for making gears is its cleanliness—the absence of impurities.

Alloy steel for barstock and forgings for gears is produced in electric furnaces, is commonly vacuum degassed, inert atmosphere shielded, and bottom poured to improve cleanliness and reduce the content of hydrogen, oxygen, and nitrogen. Improved cleanliness results in improved transverse ductility and impact strength but can reduce machinability. An example would be if sulfur content is reduced to less than .015%.

Vacuum degassed steel can be further refined by vacuum arc remelting or electroslag remelting. These processes further reduce inclusion size and content, and gas, for improved fatigue strength, to produce the highest quality steel for critical gear applications.

Two obvious "critical gear applications" are aircraft engine gearing and helicopter transmission gearing.

Cast iron is a good gear material if shock loading is minimal. When it is used, usually the driving pinion is steel, or perhaps a nonmetallic material if noise is a factor and if loads are not too great. Cast iron and the nonmetallic materials are seldom used as gear materials in the enclosed drives produced by speed reducer manufacturers.

Cast iron is a generic term for the family of high-carbon and silicon iron alloys, classified by the following categories:

Gray iron contains carbon—usually over about 3.0%—which is present as graphite flakes. Minimum hardness requirements are listed in Table 7.3.

Ductile iron, sometimes referred to as nodular iron, is characterized by the spheroidal shape of the graphite in the metal. This is produced by the introduction of magnesium and rare earth elements. Table 7.4 lists the mechanical properties of ductile iron.

Table 7.3. Minimum Hardness and Tensile Strengths for Gray Cast Iron

ASTM[a] Class Number	Brinell Hardness Number	Tensile strength (ksi)
20	155	20
30	180	30
35	205	35
40	220	40
50	250	50
60	285	60

[a]See ASTM A48 for additional information.

Table 7.4. Mechanical Properties of Ductile Iron

ASTM Grade[a]	Heat treatment	Brinell Hardness	Tensile Strength (ksi)	Yield Strength min. (ksi)	Elong. in 2 inches, min. %
60-40-18	Anneal	170 max.	60	40	18.0
65-45-12	As cast or anneal	156–207	65	45	12.0
80-55-06	Normalize	187–255	80	55	6.0
100-70-03	Quench and temper	241–302	100	70	3.0
120-90-02	Quench and temper	As specified	120	90	2.0

[a]See ASTM A536 or SAE J434 for further information.

Powdered metal parts are formed by compressing metal powders in a die cavity and heating (sintering) it to bond the particles metallurgically. Die costs are high so large quantities are necessary to amortize die costs. The powdered metal usually specified for gear applications is alloy steel.

Gear bronzes are used mostly for worm gears and are selected for their wear resistance when running with a hardened and ground steel worm. These are the phosphor or tin bronzes, manganese bronzes, aluminum bronze, and silicon bronzes. The bearing characteristics of aluminum bronze are inferior to those of the phosphor bronzes but better than for manganese bronze. The main use of silicon bronze is for lightly loaded gearing.

There are many other bronze alloys but relatively few are suitable for worm gear application. In addition to those listed above there is one that works well for many applications. It is a copper–zinc–manganese alloy, leaded, available as drop forgings or in bars. The most common of the cast bronzes is phosphor bronze, SAE 65 (now SAE C90700), which is improved if about 1% nickel is added and chill cast, or centrifugally cast if the configuration permits.

For best results in worm gear drives the worm must be carburized and hardened, with the threads ground. Also, as noted previously, for optimum results worm gearing should be run in after assembly, at no load to start, increasing to full load after a reasonable time, or until a polished area appears on the gear teeth.

Nonmetallic gear materials. There are several of these, the most common of which are the phenolic resins. One of these is canvas or linen impregnated bakelite. Usually gears made from this material are machined. This material, called *thermosetting laminates,* is available in sheets of various thicknesses, square or round rods, and tubes. It is produced by coating the fabric sheets with liquid phenol–formaldehyde resin, cutting and stacking between metal plates, and bonding under pressures of 1000 to 2500 psi and temperatures of 270° to 350° F. One common use for this material is in automobile timing gears.

The nylons, a family of *thermoplastic polymers,* usually do not have fabric reinforcing materials but can be compounded with them. Gears made from this material are either machined or made with molded teeth if the quantities are large enough to justify the cost of the molds.

In drives that must operate quietly and are not too heavily loaded, the pinion is often specified to be nonmetallic. As noted previously, tooth deflection under load is greater than for metal pinions, so such pinions should be modified oversize (long addendum) to minimize the tendency of the gear teeth to gouge into the flanks of the pinion teeth as the latter deflect.

Inspection of Materials

Hardness Testing

Materials and finished gears in the machinable hardness range are usually checked on a Brinell hardness machine, which presses a hardened steel ball into the surface to be checked. The diameter of the impression is a measure of the hardness, and is calibrated to give what is called the *Brinell Hardness Number* (BHN).

A Rockwell test machine can also be used to check materials in the machinable hardness range. This machine forces a diamond penetrator into the surface at given loads, with the depth of penetration calibrated to give the *Rockwell Hardness Number*. For hardnesses over about 400 BHN a Rockwell machine must be used. Rockwell C Scale (RC) is the most common, for which the load is 150 kg and the hardness range is about 25 to 68 RC. Other scales, loads, and ranges are as follows:

Scale	Load	Use
Rockwell A (RA)	60 kg	Small gears, medium shallow case
Rockwell 30-N (R30-N)	30 kg	Shallow depth case-hardened parts
Rockwell 15-N (R15-N)	15 kg	Very shallow depth case parts

Another device for checking hardness is the Scleroscope. This drops a hardened ball from a given height onto the surface being checked. The height of rebound is a measure of the surface hardness.

For all hardness checks the surface must be relatively smooth (say 64 microinch max.) and flat. Table 7.6 describes the various levels of surface finish, from "Extremely rough" to "Mirrorlike." The relationship between Brinell and Rockwell hardnesses and tensile strength can be obtained from steel warehouse catalogs. Table 7.5 is one of these tables, reproduced from a Central Steel & Wire Co. catalog.

Another device for checking hardness is the *Tukon*. This is a laboratory instrument used for determining case depths in test bars placed in the carburizing furnace with the gears being carburized and hardened. The hardened test bar is sectioned, the section polished to a mirror finish, and readings taken from the surface inward every .002″. More accurate results are obtained by sectioning a few of the gears from each batch after hardening if the gears are small enough, or for larger and heavier pitch gears, sectioning individual gear teeth, then polishing the sectioned areas and taking readings as above. Since this is a destructive test it is practical only when the quantities are large enough to warrant it.

Illustrations (schematic) showing sectioned gear teeth, one a carburized case pattern and one an induction hardened (by scanning) case pattern, are shown as Figures 7.1 and 7.2. By etching the polished surfaces the hardened area can be plainly seen.

Table 7.5. Hardness Conversion Table for Carbon and Alloy Steels (All Values Are Approximate)

Brinell Hardness Number (Carbide ball)	Rockwell Hardness Numbers					Tensile strength	
	C Scale	A Scale	15N Scale Superficial	B Scale	30T Scale Superficial	ksi	MPa
—	66	84.5	92.5	—	—	—	—
722	64	83.4	91.8	—	—	—	—
688	62	82.3	91.1	—	—	—	—
654	60	81.2	90.2	—	—	—	—
615	58	80.1	89.3	—	—	—	—
577	56	79.0	88.3	—	—	313	2160
543	54	78.0	87.4	—	—	292	2010
512	52	76.8	86.4	—	—	273	1880
481	50	75.9	85.5	—	—	255	1760
455	48	74.7	84.5	—	—	238	1640
443	47	74.1	83.9	—	—	229	1580
432	46	73.6	83.5	—	—	221	1520
421	45	73.1	83.0	—	—	215	1480
409	44	72.5	82.5	—	—	208	1430
400	43	72.0	82.0	—	—	201	1390
390	42	71.5	81.5	—	—	194	1340
381	41	70.9	80.9	—	—	188	1300
371	40	70.4	80.4	—	—	182	1250
362	39	69.9	79.9	—	—	177	1220
353	38	69.4	79.4	—	—	171	1180
344	37	68.9	78.8	—	—	166	1140
336	36	68.4	78.3	—	—	161	1110
327	35	67.9	77.7	—	—	156	1080
319	34	67.4	77.2	—	—	153	1050
311	33	66.8	76.6	—	—	149	1030
301	32	66.3	76.1	—	—	146	1010
294	31	65.8	75.6	—	—	141	970
286	30	65.3	75.0	—	—	138	950
279	29	64.6	74.5	—	—	135	930
271	28	64.3	73.9	—	—	131	900
264	27	63.8	73.3	—	—	128	880
258	26	63.3	72.8	—	—	125	860
253	25	62.8	72.2	—	—	123	850
247	24	62.4	71.6	—	—	119	820
243	23	62.0	71.0	—	—	117	810
240	—	—	—	100	83.1	116	800
234	—	—	—	99	82.5	114	785

continued

Table 7.5. continued

Brinell Hardness Number (Carbide ball)	Rockwell Hardness Numbers					Tensile strength	
	C Scale	A Scale	15N Scale Superficial	B Scale	30T Scale Superficial	ksi	MPa
222	—	—	—	97	81.1	104	715
210	—	—	—	95	79.8	100	690
200	—	—	—	93	78.4	94	650
195	—	—	—	92	77.8	92	635
185	—	—	—	90	76.4	89	615
176	—	—	—	88	75.1	86	590
169	—	—	—	86	73.8	83	570
162	—	—	—	84	72.4	81	560
156	—	—	—	82	71.1	77	530
150	—	—	—	80	69.7	72	495
144	—	—	—	78	68.4	69	475
139	—	—	—	76	67.1	67	460
135	—	—	—	74	65.7	65	450
130	—	—	—	72	64.4	63	435
125	—	—	—	70	63.1	61	420
121	—	—	—	68	61.7	59	405
117	—	—	—	66	60.4	57	395
114	—	—	—	64	59.0	—	—

The Tukon instrument can be used to check any hardness, so it is used also to determine core hardness of case-hardened parts, induction as well as carburized, and in determining hardness penetration in through-hardening steels.

The effective carburized case depth is usually specified to be about one sixth the tooth thickness at the pitch line and is the depth from the surface to where the hardness is 50 RC.

Inspection for Internal Defects—Ultrasonic

Ultrasonic inspection is a nondestructive test method to determine the internal soundness and cleanliness of material by passing sound (ultrasound) through it. Very short sound waves of a frequency greater than 20,000 cycles per second (audible limit) are voltage generated

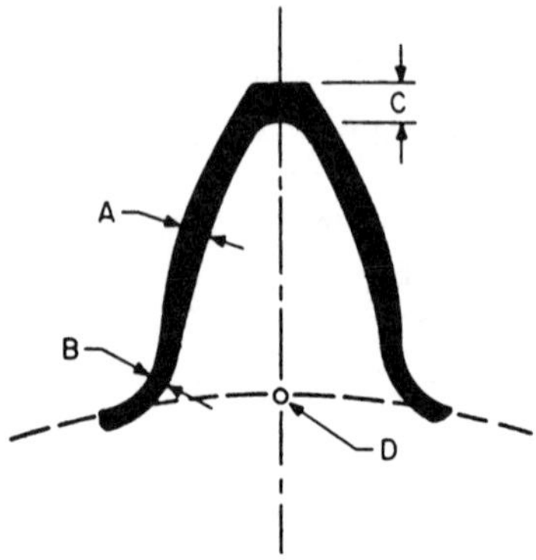

Figure 7.1. Carburized case pattern. Shaded area is all 50 RC in hardness or higher. Unshaded area is less than 50 RC. A, Profile case thickness; B, root case thickness; C, tip case thickness; D, core hardness taken here.

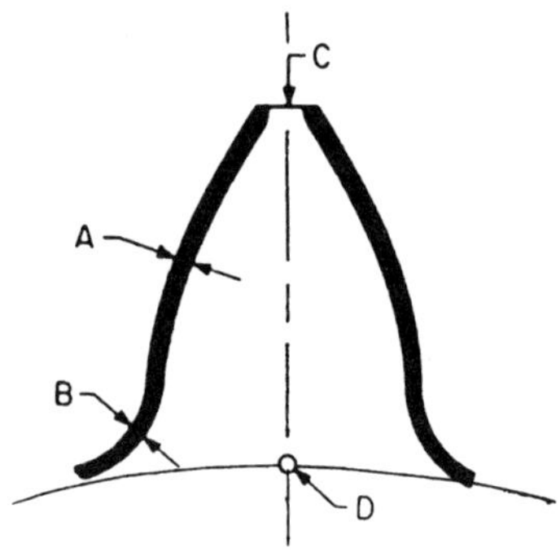

Figure 7.2. Induction hardened case pattern, by scanning method of heating. A, Profile case thickness; B, root case thickness; C, top land, no case at center; D, core hardness taken here.

and transmitted into the part by a transducer. In the method most often used, returning sound waves are transformed into voltage and monitored on an oscilloscope screen.*

---

*This statement was reproduced from ANSI/AGMA Standard 2004-B89, with permission of AGMA.

Table 7.6. Surface Finish

Symbol	Type of finish
$\overset{1000}{\surd}$	Extremely rough, like a sand casting
$\overset{500}{\surd}$	Very rough, not suitable for mating surfaces
$\overset{250}{\surd}$	Rough, heavy toolmarks
$\overset{125}{\surd}$	Fine, even toolmarks
$\overset{64}{\surd}$	Fine, no objectionable toolmarks
$\overset{32}{\surd}$	Smooth, toolmarks barely visible
$\overset{16}{\surd}$	Very smooth, ground finish
$\overset{8}{\surd}$	Near mirrorlike, polished surface
$\overset{4}{\surd}$	Mirrorlike, no marks

Surface finish requirements are designated by the symbol $\overset{xx}{\surd}$, with the numerical value *xx* in microinches.

Inspection for Surface Defects

1. *Dye penetrant.* A red dye is applied to the surface, wiped off, then a white powder in suspension is applied. If there are surface cracks the dye will have penetrated into them and will bleed through the white coating.

2. *Magnetic particle.* Magnetic particle inspection is a nondestructive testing method for locating surface and near-surface discontinuities in ferromagnetic material. When a magnetic field is introduced into the part, discontinuities lying approximately transverse to the magnetic field will cause a leakage field. Finely divided ferromagnetic particles, dry or in a water or oil suspension, are applied over the surface of the material under test. These particles will gather and hold at the leakage field, making the discontinuities visible to the naked eye.

8

Heat Treating of Steel Gears

Almost all steel gears require some sort of heat treatment, either before or after machining. An explanation of what happens when steel is heated and then cooled follows.

First it should be said that the amount of carbon present determines the steel's response to heat treatment more than any other element, in alloy as well as carbon steels. In alloy steels the various alloying elements (manganese, nickel, chromium, vanadium, and molybdenum) are added in various combinations to improve core properties in carburizing grades, and hardness depth and toughness in through-treating grades of steel. Relatively small amounts of phosphorous, silicon, and sulfur appear in all steels.

Through-Hardening Steels

We will talk first about heat treating the through-hardening grades of steel. These usually are considered to be steels that have .30% to about .50% carbon—the medium carbon range.

To start, the unmachined steel, semifinished or finished gear, or machine part is heated to what is called the critical range of temperatures. This is usually 1450° to 1650° F, which is varied with the carbon content and the alloying elements in the steel. Some examples are:

Steel	Critical range (degrees F)
1030	1575 to 1650
1045	1450 to 1550
4140	1525 to 1625
4340	1475 to 1525

The steel, gears, or parts are held at the critical temperature long enough to ensure that they are heated throughout, then cooled rapidly (quenched) by submerging in water or oil. Water is a more severe quench medium (faster cooling) than oil and is used in quenching the carbon steels. Oil must be used in quenching the alloy steels as they will harden at a slower cooling rate and would be likely to crack if quenched in water.

To explain what happens when steel is heated and then quenched we must look at its structure, which can be likened to crystalline rocks composed of different minerals. These mineral-like constituents that may occur in steel are ferrite, cementite, austenite, martensite, troostite, sorbite, and pearlite, which can be identified with the aid of a microscope after specimens have been highly polished and then etched with dilute acids that attack, in varying degree, the different constituents. These constituents are defined as follows:

Ferrite is composed largely of pure iron with small amounts of other elements in solution in the solid state. It is soft and ductile, is strongly magnetic, and has a tensile strength of about 40,000 psi and an elongation of about 40%. In its free state, ferrite is found in all carbon steels with less than .85% carbon when these steels have been cooled slowly from temperatures above or within the critical range.

Cementite is a chemical compound having the formula Fe_3C, containing 93.333% iron and 6.666% carbon by weight. Other elements also combine with carbon to form cementite. This is the hardest of all constituents occurring in steel. It is found normally in all slowly cooled steels containing over .85% carbon.

Austenite is a solid solution of carbon in iron produced at temperatures at or above the critical range. It is transformed into other constituents as the steel is cooled through the critical range. It is nonmagnetic and can be detected because of this property. *Retained austenite* in hardened steel causes the surface hardness, in the as-quenched condition, to be less than expected. It can be transformed into martensite (the hard constituent) by deep freezing, from room temperatures, by submerging in a dry ice and alcohol mixture. If this does not work, other reasons for a too-soft surface could be decarburization, deteriorating quenchant, or malfunctioning quenchant agitators.

Martensite is the hard constituent, obtained when steel is cooled rapidly from within or above the critical range. The lower the carbon content of the steel the more rapidly it must be cooled to obtain martensite. The hardness of martensite increases with the carbon content.

Troostite is formed by cooling less rapidly than to obtain martensite, or by reheating steel in which martensite has occurred. Steels that contain troostite are difficult to machine.

Sorbite is a mixture of small unoriented particles of ferrite and cementite. It is produced by regulating the rate of cooling from temperatures above the critical range or by tempering steel after hardening.

Pearlite is a mixture of ferrite and cementite in alternate layers. All steels contain some pearlite after cooling slowly from within or above the critical range. Normally pearlite has a tensile strength of about 125,000 psi and elongation of about 10%.

Figure 8.1 shows the temperatures at which these various constituents are formed, and the schematic appearance of each when etched and examined under a microscope.

Principal Heat Treatment Processes

Annealing is accomplished by heating to the critical range of temperatures and cooling slowly, usually in the furnace. The constituents present in annealed steel are ferrite and pearlite, the proportion of ferrite to pearlite increasing as the percentage of carbon increases (they are about equal at .45% carbon). The purposes of annealing are primarily to improve machinability, relieve internal stresses, and to refine the crystalline structure.

Normalizing is accomplished by heating to temperatures to about 100° F above the critical range and cooling in still air at room temperature (say about 70° F). The resultant structure is composed largely of sorbite. The purposes of normalizing are primarily to refine the grain structure prior to hardening, produce a homogeneous structure throughout, and reduce segregation in castings and forgings.

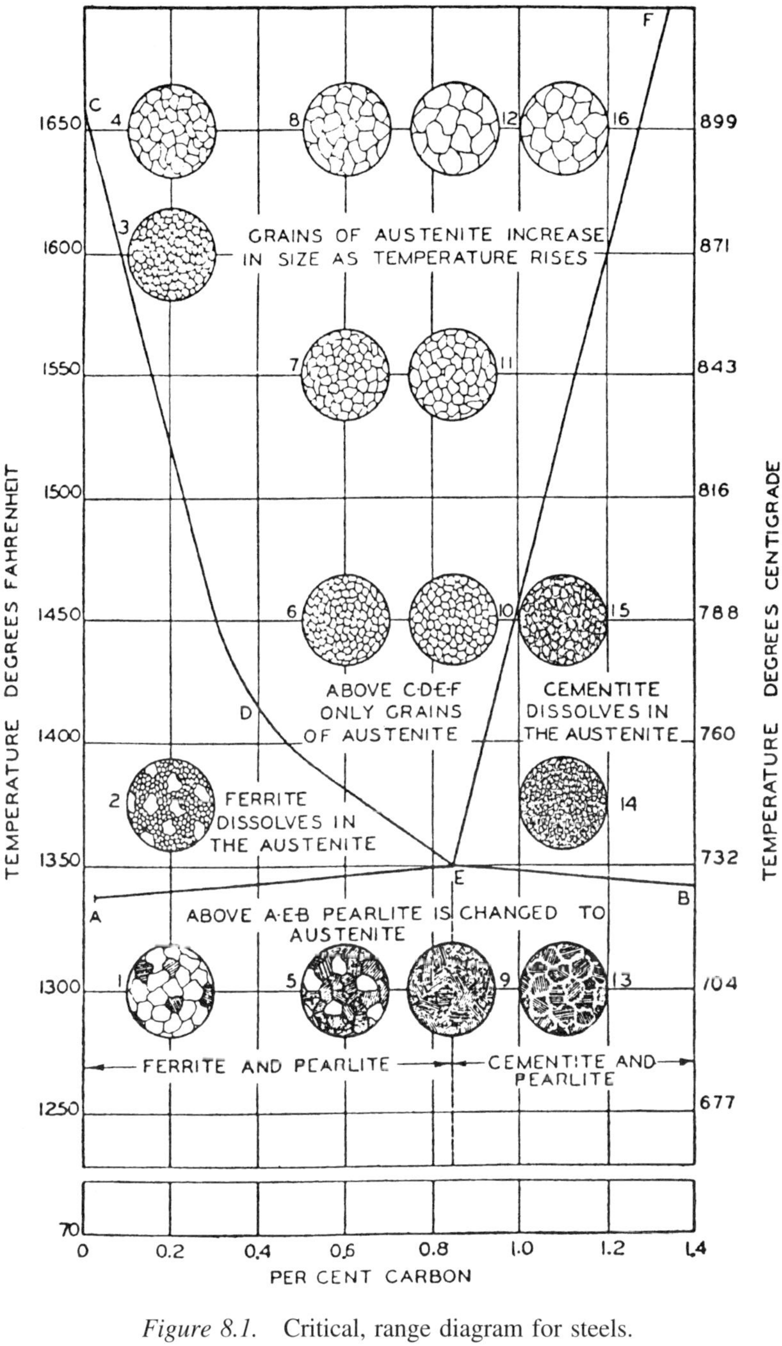

Figure 8.1. Critical, range diagram for steels.

Tempering requires reheating after hardening to temperatures below the critical range, followed by the desired rate of cooling. The purpose of tempering is to relieve stresses caused by quenching and to reduce the hardness and brittleness that result from quenching steels containing more than about .40% carbon. The higher the tempering temperature (probably a maximum of about 100° F below the critical) the lower will be the final hardness. So tempering temperature is specified according to the final hardness required.

Figure 8.2 indicates the various temperature ranges we have discussed for hardening, annealing, normalizing, and tempering.

Case Hardening

First we will discuss *carburizing and hardening*. This is achieved by heating the low-carbon steels in a carbon-rich atmosphere, most often at temperatures of 1650° to 1750° F. The steel absorbs carbon to a depth determined by the time it is in the furnace. Steels for carburizing usually contain 0.15% to 0.20% carbon. After carburizing the case will contain about 1.0% carbon at the surface, diminishing to 0.40% to 0.45% carbon at the depth where a hardness of 50 RC is required. This is known as the effective case thickness (see "Inspection of Materials," Tukon inspection procedure).

Some of the steels frequently requiring carburizing and hardening are 1015, 1020, 8620, and 4320. The first two are carbon steels, the second two alloys. The core properties of the carbon steels will be affected very little by the carburizing and hardening process, but the core properties of the alloys will be improved as they have hardenability characteristics of their own. This is why alloy low-carbon steels are selected for applications requiring tooth strength as well as tooth surface hardness.

To harden, the parts are quenched from the carburizing temperature, then for best results are reheated to 1400° to 1450° F and again quenched. If the surface hardness is less than expected, retained austenite could again be the culprit, which as stated for through-hardening steels can be eliminated by deep-freezing in a mixture of dry ice and alcohol.

Nitriding is another case hardening process. The steels most often used are a special alloy called *Nitralloy,* usually containing about .40% carbon and pretreated to the desired core properties. Other alloys such

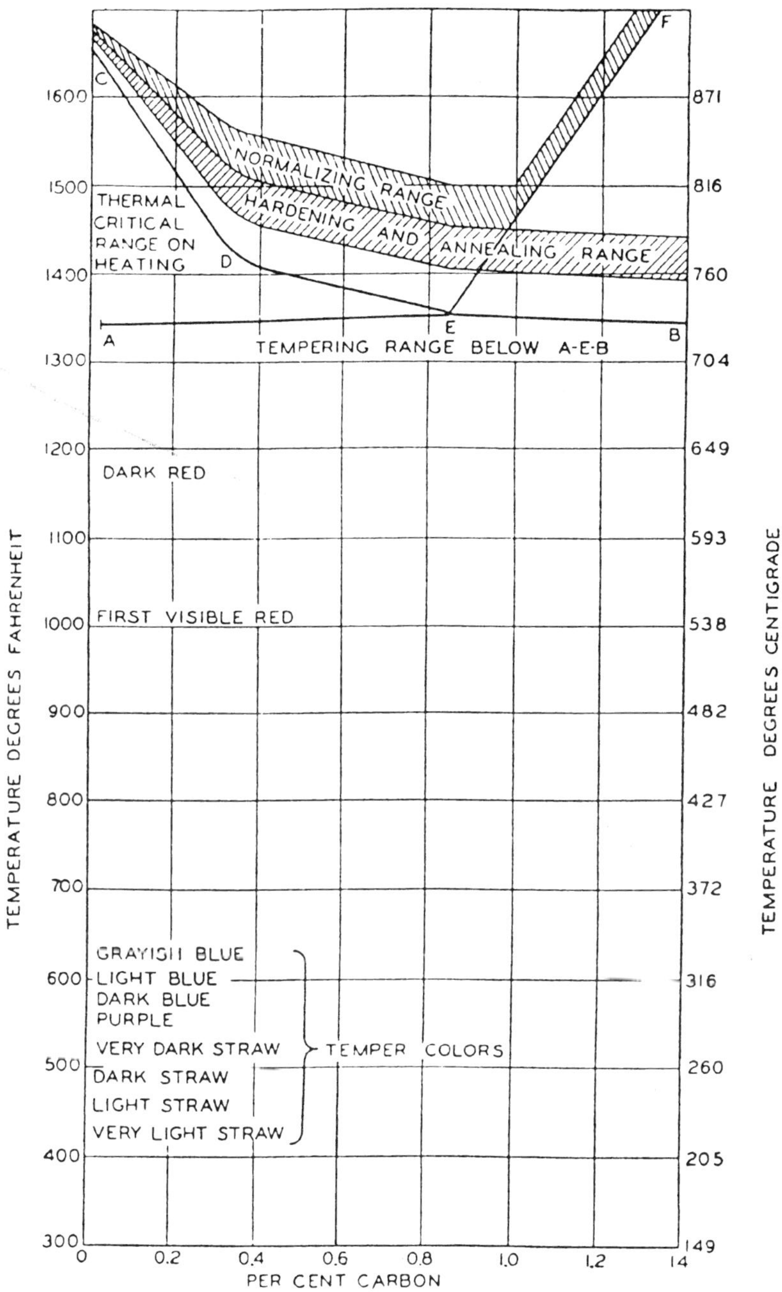

Figure 8.2. Temperature ranges used in heat treating steels.

as 4140 can also be nitrided. This process is accomplished by heating to relatively low temperatures (about 950° F) in an atmosphere of dry ammonia gas for 10 to 40 hours depending on the size of the part and required case depth, then cooled in air. The case is harder than obtained with carburizing and hardening, and much shallower. Because of the low temperature used in nitriding and because the part is not quenched, distortion is less than occurs in carburizing and hardening.

Flame Hardening is a case hardening process when tooth-to-tooth hardening is used for the heavier pitches. Through-hardening steels are required, so when finer pitch gears are spin hardened the teeth are hardened throughout.

In tooth-to-tooth flame hardening an oxyfuel burner, or flame head, is used, shaped to heat the flanks of the tooth. Theoretically the flame head can be shaped to harden the roots as well, but it is difficult to achieve consistency in root hardness and depth. The flame head is followed by a quench spray nozzle as the flame head traverses the length of each tooth. The quench medium is usually water.

Induction hardening of through-hardening steels, tooth-to-tooth, heats the tooth flanks, and the roots when required, by inducing an electric current into the surfaces to be hardened. This is accomplished by passing high-frequency alternating electric current through an inductor block, which in turn produces a high-frequency magnetic field that induces a current in the surfaces of the teeth, heating them to the critical temperature range. The depth of heating is controlled by varying the speed of scanning, the frequency, and the power. The inductor is followed by a quench spray nozzle having a shape similar to that of the inductor. The quench medium is water with an additive to reduce the severity of the quench to approximate an oil quench. Oil cannot be used, as it would clog the spray nozzle and because of danger of fire. The inductor, which must not touch the workpiece surfaces, is shaped to harden the flanks of each tooth when the roots are not to be hardened, or the flanks of adjacent teeth *and* the root and fillets when they are. The latter is recommended as tooth bending strength is increased. *It is important* to make up a two- or three-tooth section, harden it by scanning, then check to make sure the depth, pattern, and hardness are as specified before hardening a complete gear.

Finer pitch gears—finer than about 5 *DP*—are spin induction hardened by use of a wraparound coil. This hardens the teeth throughout.

Illustrations of the various types of coils, inductor blocks, or flame heads used, and hardness patterns obtained, are shown schematically in Figure 8.3.

Induction hardening is preferred over flame hardening because hardness and depth are more accurately controlled.

There are induction hardening machines available that do their work

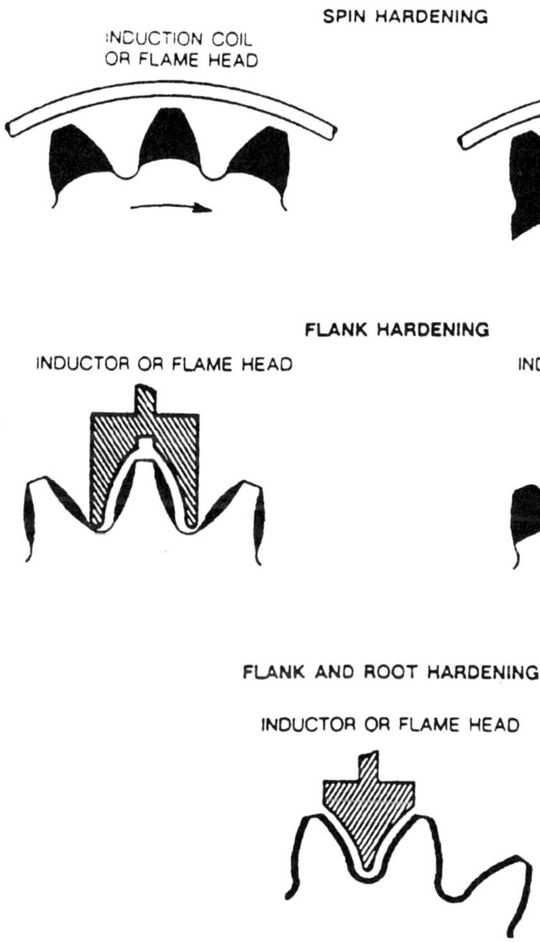

Figure 8.3. Various coils, inductors, and flame heads used, and hardness patterns obtained, when flame or induction hardening. (From ANSI/AGMA 2004-B89) with permission by AGMA.

with the gears being hardened submerged in water, or water with an additive. It is said that distortion is less when such machines are used.

Flame or induction hardening are specified when the size and configuration of the gear make it impractical to carburize and harden.

Sometimes it is necessary to case harden the teeth only, leaving the rest of the gear soft for finish machining after the teeth are hardened. This of course is accomplished in induction or flame hardening. It is accomplished in carburizing by painting the surfaces to remain soft with a copper powder suspended paint or by copper plating (which requires masking the teeth). The painting or plating prevents carbon penetration in the painted or plated surfaces.

A recent development in induction hardening is called the Micropulse® System, developed by Contour Hardening, Inc. This system will rapidly harden gear teeth and other irregular shapes that require a high degree of accuracy.

In this system a gear is preheated by radiofrequency pulses, then given a final pulse to heat it to the required critical temperature, from which it is quenched by a water-base solution applied through the coil. It is said that the cycle time is fast and that distortion is minimal.

Contour Hardening, Inc. is located in Indianapolis, IN.

Shot peening is the bombardment of stress concentration surfaces, such as the roots and fillets of gear teeth, with hardened steel or glass shot. The diameters of the shot used are related to the fillet radius of the gear teeth to be shot peened. Usually the flanks of the teeth are not directly peened. Shot peening produces minute indentations in the surfaces at which the peening device is aimed. The pellets are propelled by centrifugal force from a rotating wheel or by air pressure. These indentations produce residual compressive stress in the peened surfaces. This increases the bending strength of gear teeth as under load the usual tensile stress from bending has to first overcome the compressive stress before the teeth start to bend. The result is that shot peened teeth can have as much as 25% greater strength, both bending and fatigue, than teeth that have not been shot peened.

Probably the greatest benefit from shot peening is for hardened gears, as the quenching of these gears can cause surface tensile stresses that shot peening can overcome. It is important that any heat treating be done prior to shot peening as heating after peening will draw the peened surfaces, eliminating the desired residual compressive stress.

It is recommended that gears which are expected to be highly stressed be subjected to shot peening, with automated equipment, operated by experienced people, in shops that have adequate checking capabilities.

Shot peening is not to be confused with blasting for cleaning.

Corrosion

Except for the stainless steels all steels are subject to corrosion, which is the formation of iron oxide, or rust. This can be minimized by plating with various metals. Some of the most common are zinc, cadmium, and chromium. Copper plating is also used, but primarily as a stop-off when carburizing parts, rather than as a rust preventative.

Before specifying any of the above, or other, rust preventive methods, the designer should familiarize himself with plating practices.

9

Splines, Keyways, and Keys

Splines are a series of teeth, or slots and keys, on a shaft and in the bore of a gear, sprocket, sheave, or any rotating part that must be secured to a shaft to transmit motion and power from one to the other. Single and multiple keyways and keys serve the same purpose.

There are two types of splines, involute and parallel side. We will discuss these first, to be followed by keyways and keys.

Involute Splines

Involute splines have the same form as involute gear teeth except the teeth are one half the depth of standard full-depth gear teeth and the pressure angle is 30°. There are also 37.5° and 45° PA standard involute splines but these are less common than the 30°.

The involute spline is designated by a fraction, similar to the Fellows stub for gear teeth, in which the numerator is the diametral pitch and the denominator denotes the tooth depth and is always twice the numerator. There are 14 standard pitches, as follows: 2.5/5, 3/6, 4/8, 5/10, 6/12, 8/16, 10/20, 12/24, 16/32, 20/40, 24/48, 32/64, 40/80, and 48/96. The minimum number of teeth is 6 and the maximum is 60. There are both flat root and fillet root types, illustrated in Figures 9.1 and 9.2.

There are three types of fits: (1) *major diameter,* a fit controlled by varying the major diameter of the external spline; (2) *sides of teeth,* a fit controlled by varying the tooth thickness, customarily used for fillet

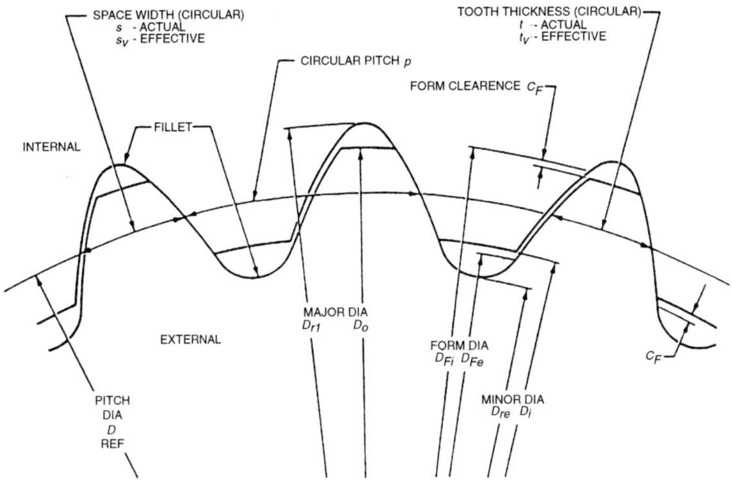

Figure 9.1. Involute spline, fillet root, side fit.

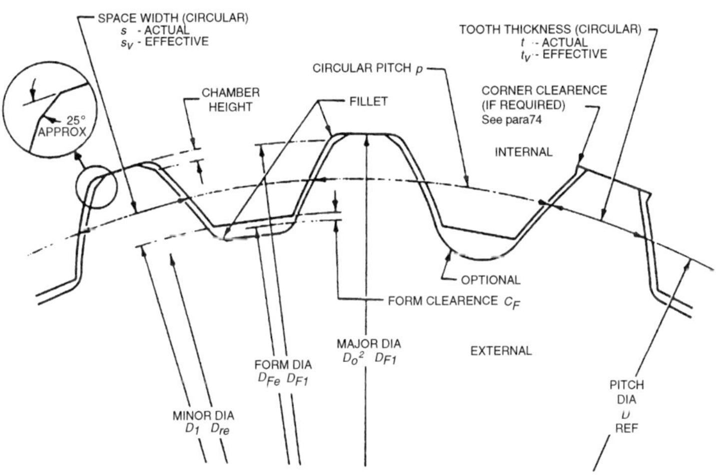

Figure 9.2. Involute spline, flat root, major diameter fit.

root splines; and (3) *minor diameter,* a fit controlled by varying the minor diameter of the external spline or of the internal spline by grinding the bore after hardening or by quenching on plugs to prevent shrinkage.

Each type of fit is further divided into three classes: (a) *sliding,* clearance at all points; (b) *close,* close on either major or minor diameter or sides of teeth; and (c) *press,* interference on either the major or minor diameter or sides of teeth. In all cases the external spline dimensions are varied to control the fit, as the internal spline is often broached and cannot be varied except by finishing the minor diameter after hardening.

Involute spline proportions, dimensions, fits, and tolerances are given in detail in the SAE publication "Involute Splines and Inspection, ANSI B92.1-1970," 7th printing, September, 1988, from which the foregoing data were taken. This publication, however, does not provide data for minor diameter fits. Such are necessary in certain conditions, such as when the internal spline is in a hardened gear and the minor diameter of the spline is to be ground after hardening. Then in hobbing the external, which must be flat root, the depth of cut is varied to fit the minor diameter of the external spline to the finished minor diameter of the internal spline.

Some of the advantages of involute over parallel side splines are that standard gear inspection procedures can be used, they are self centering, which equalizes stresses, and they have maximum strength at the roots of the teeth.

Parallel Side Splines

SAE Standard parallel side splines are illustrated in Figure 9.3 and Tables 9.1 and 9.2, with dimensions and proportions. They are equally spaced keyways. External splines can be milled on a milling machine equipped with an indexing mechanism, or hobbed if the special hobs are available. Internal parallel side splines can be cut on a keyseater equipped with an indexing mechanism, or broached if broaches are available.

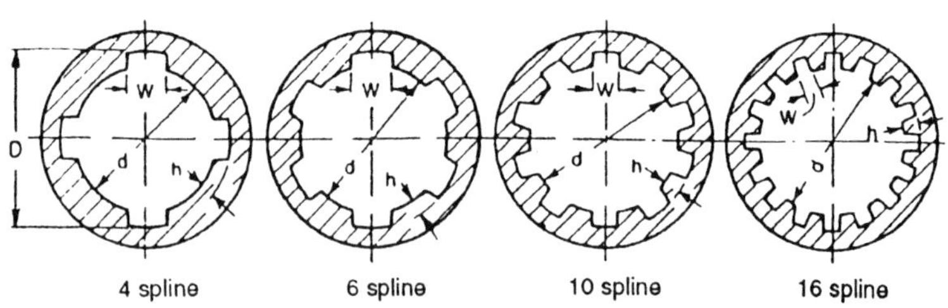

Figure 9.3. Parallel-side splines.

Table 9.1. Dimensions of Spline Fittings (SAF, Standard. All Dimensions in Inches)

Nominal diam	4-Spline for all fits		6-Spline for all fits		10-Spline for all fits		16-Spline for all fits	
	D_{max}[a]	W_{max}[b]	D_{max}[a]	W_{max}[b]	D_{max}[a]	W_{max}[b]	D_{max}[a]	W_{max}[b]
3/4	0.750	0.181	0.750	0.188	0.750	0.117		
7/8	0.875	0.211	0.875	0.219	0.875	0.137		
1	1.000	0.241	1.000	0.250	1.000	0.156		
1 1/8	1.125	0.271	1.125	0.281	1.125	0.176		
1 1/4	1.250	0.301	1.250	0.313	1.250	0.195		
1 3/8	1.375	0.331	1.375	0.344	1.375	0.215		
1 1/2	1.500	0.361	1.500	0.375	1.500	0.234		
1 5/8	1.625	0.391	1.625	0.406	1.625	0.254		
1 3/4	1.750	0.422	1.750	0.438	1.750	0.273		
2	2.000	0 482	2.000	0.500	2.000	0.312	2.000	0.196
2 1/4	2.250	0.542	0.250	0.563	2.250	0.351		
2 1/2	2.500	0.602	2.500	0.625	2.500	0.390	2.500	0.245
3	3.000	0.723	3.000	0.750	3.000	0.468	3.000	0.294
3 1/2	—	—	—	—	3.500	0.546	3.500	0.343
4	—	—	—	—	4.000	0.624	4.000	0.392
4 1/2	—	—	—	—	4.500	0.702	4.500	0.441
5	—	—	—	—	5.000	0.780	5.000	0.490
5 1/2	—	—	—	—	5.500	0.858	5.500	0.539
6	—	—	—	—	6.000	0.936	6.000	0.588

[a]Tolerance allowed of –0.001 in. for shafts 3/4 to 1 3/4 inches inclusive; –0.002 for shafts 2 to 3 inches, inclusive; –0.003 inches for shafts 3 1/2 to 6 inches inclusive, for 4-, 6-, and 10-spline fittings; tolerance of –0.003 inch allowed for all sizes of 16-spline fittings.

[b]Tolerance allowed of –0.002 inch for shafts 3/4 inch to 1 3/4 inch inclusive; of –0.0003 inch for shafts 2 to 6 inches inclusive, for 4-, 6-, and 10-spline fittings; tolerance of –0.003 allowed for all sizes of 16-spline fittings.

Table 9.2. Spline Proportions

No. of splines	*W* for all fits	Permanent fit		To slide when not under load		To slide under load	
		h	*d*	*h*	*d*	*h*	*d*
4	0.241*D*	0.075*D*	0.850*D*	0.125*D*	0.750*D*		
6	0.250*D*	0.050*D*	0.900*D*	0.075*D*	0.850*D*	0.100*D*	0.800*D*
10	0.156*D*	0.045*D*	0.910*D*	0.070*D*	0.860*D*	0.095*D*	0.810*D*
16	0.098*D*	0.045*D*	0.910*D*	0.070*D*	0.860*D*	0.095*D*	0.810*D*

Keys and Keyways

As stated previously, keyways are produced in the bore of the gear by broaching or shaped in a keyseater. They are produced in the shaft by milling in a milling machine.

There are three basic types of keys: square section, Woodruff, and Gib-head keys. These are illustrated in the figures and tables that follow.

The keyways for square keys are usually dimensioned "across the bore" for the internal KW and from the bottom of the KW in the shaft to the shaft diameter 180° away, as illustrated in Figure 9.4. Standard sizes are listed in Table 9.3.

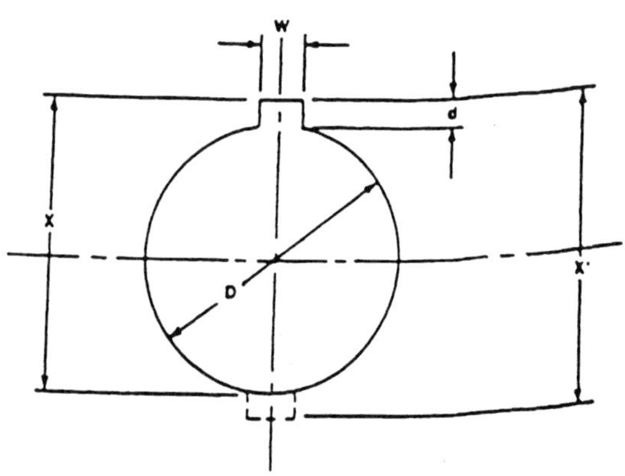

Figure 9.4. Keyways for square (to 1-inch) keys, rectangular 1¼ and 1½ inches.

Table 9.3. Standard Keyways and Setscrews for Shafts

Diameter of hole (inches)	*W*	*d*	Size of setscrew
5/16 to 7/16	3/32	3/64	10–32
1/2 to 9/16	1/8	1/16	1/4–20
5/8 to 7/8	3/16	3/32	5/16–18
15/16 to 1 1/4	1/4	1/8	3/8–16
1 5/16 to 1 3/8	5/16	5/32	7/16–14
1 7/16 to 1 3/4	3/8	3/16	1/2–13
1 13/16 to 2 1/4	1/2	1/4	9/16–12
2 5/16 to 2 3/4	5/8	5/16	5/8–11
2 13/16 to 3 1/4	3/4	3/8	3/4–10
3 5/16 to 3 3/4	7/8	7/16	7/8–9
3 13/16 to 4 1/2	1	1/2	1–8
4 9/16 to 5 1/2	1 1/4	7/16	1 1/8–7
5 9/16 to 6 1/2	1 1/2	1/2	1 1/4–6

A Woodruff key is illustrated in Figure 9.5, with dimensions listed in Table 9.4. A Gib-head key is illustrated in Figure 9.6 with dimensions listed in Table 9.5.

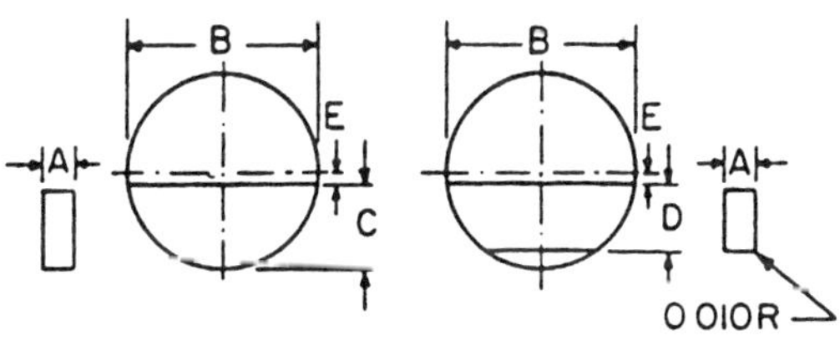

Figure 9.5. Woodruff key.

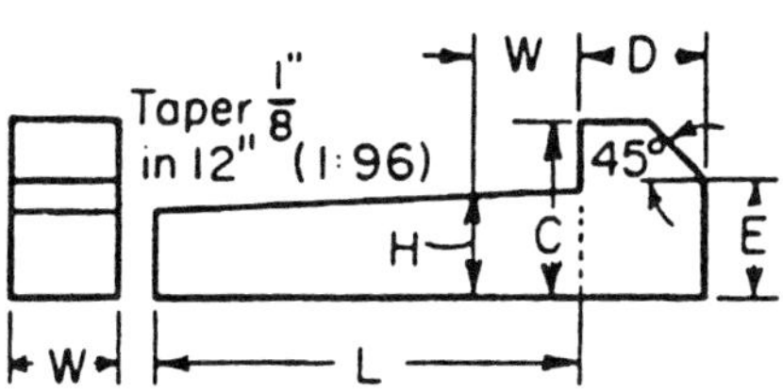

Figure 9.6. Gib-head taper stock key.

Table 9.4. Woodruff Key Dimensions [ANSI B17.2–1967 (R1978)] (All Dimensions in Inches)

Key No.	Nominal key size, $A \times B$	Width of Key, A		Diam of key, B		Height of key				Distance below E
						C		D		
		Max.	Min.	Max.	Min.	Max.	Min.	Max.	Min.	
204	1/16 × 1/2	0.0635	0.0625	0.500	0.490	0.203	0.198	0.194	0.188	3/64
304	3/32 × 1/2	0.0948	0.0928	0.500	0.490	0.203	0.198	0.194	0.188	3/64
305	3/32 × 5/8	0.0948	0.0938	0.625	0.615	0.250	0.245	0.240	0.234	1/16
404	1/8 × 1/2	0.1260	0.1250	0.500	0.490	0.203	0.198	0.194	0.188	3/64
405	1/8 × 5/8	0.1260	0.1250	0.625	0.615	0.250	0.245	0.240	0.234	1/16
406	1/8 × 3/4	0.1260	0.1250	0.750	0.740	0.313	0.308	0.303	0.297	1/16
505	5/32 × 5/8	0.1573	0.1563	0.625	0.615	0.250	0.245	0.240	0.234	1/16
506	5/32 × 3/4	0.1573	0.1563	0.750	0.740	0.313	0.308	0.303	0.297	1/16
507	5/32 × 7/8	0.1573	0.1563	0.875	0.865	0.375	0.370	0.365	0.359	1/16
606	3/16 × 3/4	0.1885	0.1875	0.750	0.740	0.313	0.308	0.303	0.297	1/16
607	3/16 × 7/8	0.1885	0.1875	0.875	0.865	0.375	0.370	0.365	0.359	1/16
608	3/16 × 1	0.1885	0.1875	1.000	0.990	0.438	0.433	0.428	0.422	1/16
609	3/16 × 1/8	0.1885	0.1875	1.125	1.115	0.484	0.479	0.475	0.469	5/64

continued

Table 9.4. continued

807	1/4 × 7/8	0.2510	0.2500	0.875	0.865	0.375	0.370	0.365	0.359	1/16
808	1/4 × 1	0.2510	0.2500	1.000	0.990	0.438	0.433	0.428	0.422	1/16
809	1/4 × 1 1/8	0.2510	0.2500	1.125	1.115	0.484	0.479	0.475	0.469	5/64
810	1/4 × 1 1/4	0.2510	0.2500	1.250	1.240	0.547	0.542	0.537	0.531	5/64
811	1/4 × 1 3/8	0.2510	0.2500	1.375	1.365	0.594	0.589	0.584	0.578	3/32
812	1/4 × 1 1/2	0.2510	0.2500	1.500	1.490	0.641	0.636	0.631	0.625	7/64
1008	5/16 × 1	0.3135	0.3125	1.000	0.990	0.438	0.433	0.428	0.422	1/16
1009	5/16 × 1 1/8	0.3135	0.3125	1.125	1.115	0.484	0.479	0.475	0.469	5/64
1010	5/16 × 1 1/4	0.3135	0.3125	1.250	1.240	0.547	0.542	0.537	0.531	5/64
1011	5/16 × 1 3/8	0.3135	0.3125	1.375	1.365	1.594	0.589	0.584	0.578	3/32
1012	5/16 × 1 1/2	0.3135	0.3125	1.500	1.490	0.641	0.636	0.631	0.625	7/64
1210	3/8 × 1 1/4	0.3760	0.3750	1.250	1.240	0.547	0.542	0.537	0.531	5/64
1211	3/8 × 1 3/8	0.3760	0.3750	1.375	1.365	0.594	0.589	0.584	0.578	3/32
1212	3/8 × 1 1/2	0.3760	0.3750	1.500	1.490	0.641	0.636	0.631	0.625	7/64

Numbers indicate the nominal key dimensions. The last two digits give the nominal diameter (*B*) in eighths of an inch and the digits preceding the last two give the nominal width (*A*) in thirty-seconds of an inch. Thus, 204 indicates a key 2/32 × 4/8 or 1/16 × 1/2 inch; 1210 indicates a key 12/32 × 10/8 or 3/8 × 1 1/4 inches.

Table 9.5. Dimensions of Square and Flat Gib-Head Taper Stock Keys[a] (All Dimensions in Inches)

	Square type					Flat type					Tolerance	
	Key		Gib-head			Key		Gib-head				
Shaft diameter	Max. width, *W*	Height at large end,[b] *H*	Height, *C*	Length, *D*	Height edge of chamfer, *E*	Max. width, *W*	Height at large end,[b] *H*	Height, *C*	Length, *D*	Height edge of chamfer, *E*	On width (−)	On height (+)
1/2 to 8/16	1/8	1/8	1/4	7/32	5/32	1/8	3/32	3/16	1/8	1/8	0.0020	0.0020
5/8 to 7/8	3/16	3/16	5/16	9/32	7/32	3/16	1/8	1/4	3/16	5/32	0.0020	0.0020
1 5/16 to 1 1/4	1/4	1/4	7/16	11/32	11/32	1/4	3/16	5/16	1/4	3/16	0.0020	0.0020
1 5/16 to 1 3/8	5/16	5/16	9/16	13/32	13/32	5/16	1/4	3/8	5/16	1/4	0.0020	0.0020
1 7/16 to 1 3/4	3/8	3/8	11/16	15/32	15/32	3/8	1/4	7/16	3/8	5/16	0.0020	0.0020
1 13/16 to 2 1/4	1/2	1/2	7/8	19/32	5/8	1/2	3/8	5/8	1/2	7/16	0.0025	0.0025
2 5/16 to 2 3/4	5/8	5/8	1 1/16	23/32	3/4	5/8	7/16	3/4	5/8	1/2	0.0025	0.0025
2 7/8 to 3 1/4	3/4	3/4	1 1/4	7/8	7/8	3/4	1/2	7/8	3/4	5/8	0.0025	0.0025
3 3/8 to 3 3/4	7/8	7/8	1 1/2	1	1	7/8	5/8	1 1/16	7/8	3/4	0.0030	0.0030
3 7/8 to 4 1/2	1	1	1 3/4	1 3/16	1 3/16	1	3/4	1 1/4	1	13/16	0.0030	0.0030
4 3/4 to 5 1/2	1 1/4	1 1/4	2	1 7/16	1 7/16	1 1/4	7/8	1 1/2	1 1/4	1	0.0030	0.0030
5 3/4 to 6	1 1/2	1 1/2	2 1/2	1 3/4	1 3/4	1 1/2	1	1 3/4	1 1/2	1 1/4	0.0030	0.0030

[a]Stock keys are applicable to the general run of work and the tolerances have been set accordingly. They are not intended to cover the finer applications where a closer fit may be required.

[b]This height of the key is measured at the distance *W* equal to the width of the key, from the gib-head.

Tables and illustrations in this chapter are reproduced from Marks' Standard Handbook for Mechanical Engineers, Ninth Edition, McGraw-Hill, New York, 1987, by permission.

10

Gear Failures

Almost every gear user will sooner or later be confronted with gear failures. Some of the most common are discussed in this chapter. Examples are illustrated in Figures 10.2 through 10.8. They are described in the following sections.

Pitting

This is a surface fatigue failure which takes a while to occur. It is designated as "micro" (small pits) or "macro" (large pits). Usually micro pitting occurs first. Pitting is caused by load concentration from misalignment or errors in profile or tooth spacing, by bad design, or by lubrication problems.

Breakage

Breakage is caused by extensive macro pitting, which weakens the teeth, or most often by fatigue, with a crack usually starting at the tooth fillet. Fatigue results from repetitive stressing and can be detected by a relatively smooth surface from where the crack started, then a jagged surface where the break occurred. Breakage from sheer overload will have an all-jagged surface at the break.

Figure 10.1. A case-hardened helical pinion with teeth broken and shattered, the result of severe surface damage from pitting. (Courtesy of Dudley Engineering Group, Inc., San Diego, CA.)

Figure 10.2. Simple bending fatigue on case-hardened spiral bevel gear teeth. Overload at heel of teeth was due to misfit with pinion. (Courtesy of Dudley Engineering Group, Inc., San Diego, CA.)

Figure 10.3. Severe pitting on case-hardened helical pinion teeth. Continued running would have resulted in tooth breakage. (Courtesy of Dudley Technical Group, Inc., San Diego, CA.)

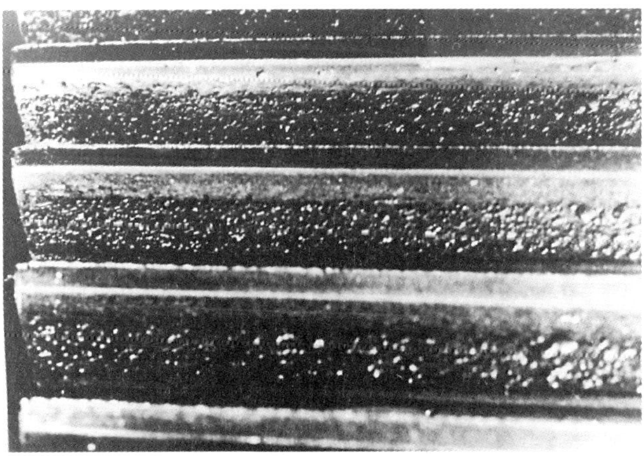

Figure 10.4. Micro and macro pitting on medium-hard helical gear teeth. The speed was not fast enough to develop a good elastohydrodynamic oil film and tooth loading was too heavy. (Courtesy of Dudley Technical Group, Inc., San Diego, CA.)

Figure 10.5. Serious "ledge" wear on medium-hard helical gear teeth due to micro and macro pitting. Tips of teeth now carry most of the load. With continued running, ultimate failure will probably be tooth breakage. (Courtesy of Dudley Technical Group, Inc., San Diego, CA.)

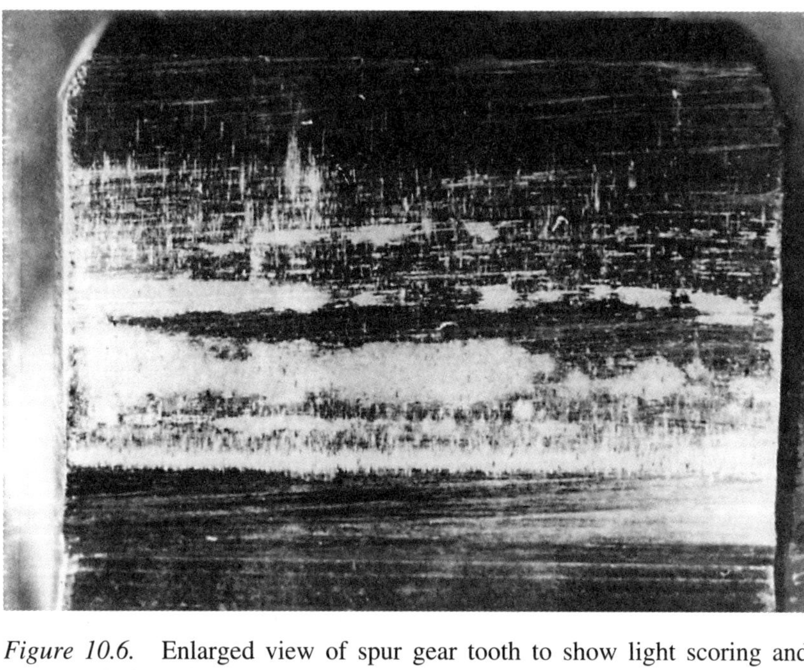

Figure 10.6. Enlarged view of spur gear tooth to show light scoring and patches of micro pitting. This gear was running hot with a thin oil and had a poor surface finish. (Courtesy of Dudley Technical Group, Inc., San Diego, CA.)

Figure 10.7 Case-hardened ground tooth helical gear with pronounced waves in the ground finish. There is micro pitting on most of the wave crests. (Courtesy of Dudley Technical Group, Inc., San Diego, CA.)

Scoring

Scoring results, usually, because there is a lubrication problem. It can result also from bad design, poor workmanship, poor surface finish, involute error, and other factors. The two degrees of scoring are called "initial scoring" and "severe scoring." Initial scoring often stops after a while and severe scoring keeps going until enough material is removed

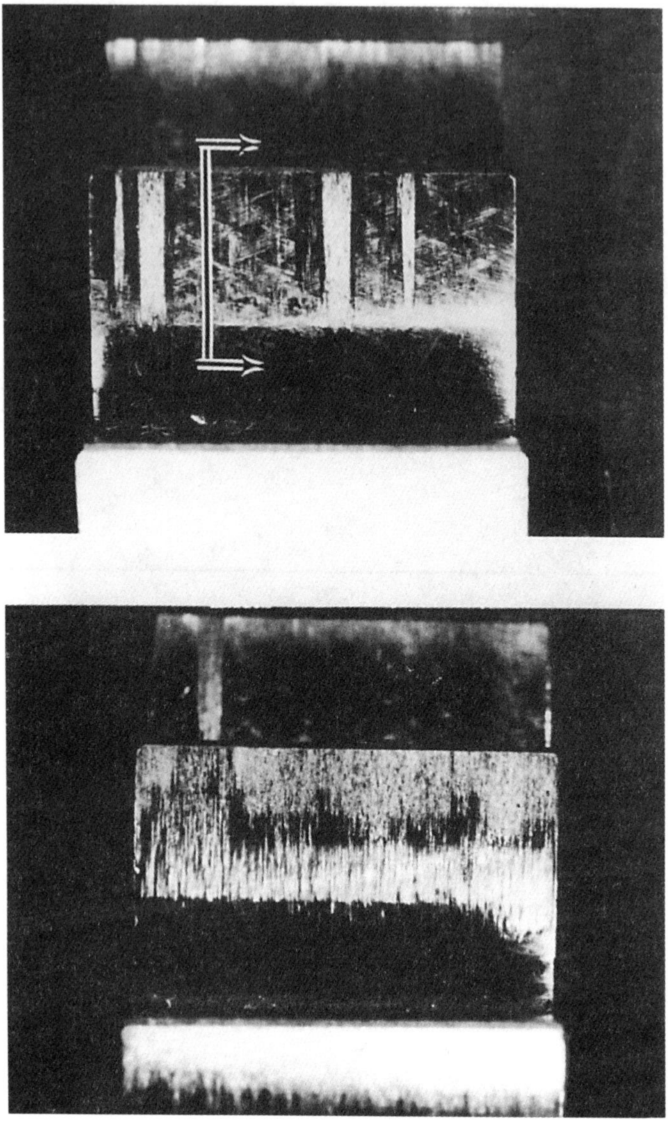

Figure 10.8. Upper view shows a close-up of a tooth flank with just four radial score marks. The lower view shows a tooth flank completely covered with scoring. These were case hardened and ground teeth, heavily loaded and run with an oil without strong chemical additives to prevent scoring. (Courtesy of Dudley Technical Group, Inc., San Diego, CA.)

to weaken the tooth, resulting in breakage. Initial scoring will be relatively smooth and severe scoring will cause a rough surface. If initial scoring does not stop it will become severe.

There are several more ways a gear tooth can be damaged and more kinds of tooth breakage. In this chapter we talk only about the most common. To determine the exact causes of a failure a gear expert should be consulted, and it is important to do this so steps can be taken to avoid trouble in the future. For example, a change in the type of lubricant could be all that is necessary, after it is first determined that misalignment, backlash, and runout are within acceptable limits, and that the hardness of pinion and gear are as specified.

If there is tooth profile error the gears probably should be replaced. Such error can be detected by observing the wear pattern on the teeth. If the wear is greater at the tip of the pinion teeth, for example, this would indicate that the pinion was cut with a hob that generated a plus involute at the tips or provided insufficient tip relief, or both.

11

Efficiency of Gears

This chapter is about gear tooth mesh and worm thread and worm gear tooth mesh efficiencies, from which the heat, or power, losses can be determined (power loss = 100 − efficiency).

The tooth mesh efficiencies and corresponding power losses of spur, helical, herringbone, straight, and spiral bevel and Zerol® gears all will average about .98 to .99 (98% to 99%). The heat or power losses are therefore about 1.0% to 2.0%. Although these power losses seem small and are often ignored by designers, they are important because they must be considered, along with other losses that occur in a drive, when looking at the amount of heat from friction that must be dissipated.

Hypoid gears, because they have higher tooth rubbing velocities than spiral bevels, will have a tooth mesh efficiency of about .90 (90%) or a power loss of about 10%.

Now we consider worm and worm gear drives. They are different because their efficiency varies with the worm's lead angle, the pressure angle, the position of the pitch plane of the worm, and other factors as well. The thread-tooth action is primarily sliding. To illustrate we will have a look at the worm and worm gear drives described as follows: the 1st—single thread, the 2nd—triple thread, and the 3rd—sextuple thread.

Worm—hardened steel with the threads ground, 3.000″ *PD*, .500″ pitch (axial), 30° PA (normal)
Worm gear—Phosphor bronze (SAE 65)

Shaft angle—90°
Worm speed—1000 rpm.

Let:

P = axial pitch, inches
L = lead, inches
R_1 = pitch radius of worm, inches
λ = lead angle
ϕ_n = normal pressure angle
V = velocity of worm at R_1, ft/min
V_s = sliding velocity, ft/min
n = rpm of worm
f = coefficient of friction.

Then in our first example:
1. Single thread (one start) (.500 ″ lead)

$$P = .500'' \quad R_1 = 1.500'' \quad L = .500'' \quad \tan\lambda = L/2 \times \pi \times R_1$$
$$= 0.500/2 \times \pi \times 1.500$$
$$= 0.053052$$
$$\lambda = 3^\circ\ 02'\ 13''$$
$$\sin 2\lambda = .10581$$
$$\cos\lambda = .99859$$
$$\cos\phi_n = .86603$$
$$V = 3.000 \times \pi \times 1000/12 = 785.4 \text{ ft/min}$$
$$V_s = V/\cos\lambda = 785.4/.99859 = 786.5 \text{ ft/min}$$
$$f = .030 \text{ (from tables—experimental value)*}$$

*The value of f varies with sliding velocity (V_s) approximately as follows:

V_s	f^2
200	0.05
400	0.04
800	0.03
1000	0.02

The values of f are experimental values for the worm and gear materials indicated herein. They were taken from Chapter 14, p. 10 of *Gear Handbook,* Darle W. Dudley, Editor, McGraw-Hill, New York, 1962. The author of chapter 14 was Eugene E. Shipley.

$$\text{Efficiency } (E)^{\dagger} = \frac{\cos \phi_n \times \sin 2\lambda}{\cos \phi_n \times \sin 2\lambda + 2f}$$
$$= \frac{.86603 \times .10581}{.86603 \times .10581 + .06} \quad (1)$$
$$= \frac{0.09163463}{0.15163463} = .60\ (60\%)$$
$$\text{Power loss} = 100 - E = 40\%$$

In our second example:

2. Same worm specs as (1) except triple thread (three starts) (1.500″ lead)

$$L = 1.500'' \qquad \tan \lambda = 1.500/2 \times \pi \times 1.500$$
$$= .15915$$
$$\lambda = 9^\circ\ 02'\ 34''$$
$$\sin 2\lambda = .31044$$
$$\cos \lambda = .98758$$
$$\cos \phi_n = .86603$$
$$V_s = V/\cos \lambda = 785.4/.98758 = 795.3 \text{ ft/min}$$
$$E = \frac{.86603 \times .31044}{.86603 \times .31044 + .06} = .82\ (82\%)$$
$$\text{Power loss} = 100 - E = 18\%$$

And in our third example:

3. Same worm specs as in (1) except sextuple thread (six starts) (3.000″ lead)

$$L = 3.000'' \qquad \tan \lambda = 3.000/2 \times \pi \times 1.500$$
$$= .31831$$
$$\lambda = 17^\circ\ 39'\ 24''$$
$$\sin 2\lambda = .57804$$
$$\cos \lambda = .95289$$
$$\cos \phi_n = .86603$$
$$V_s = V/\cos \lambda = 785.4/.95289 = 824.2 \text{ ft/min}$$

†The formula for efficiency (E) was taken from p. 416 of *Analytical Mechanics of Gears,* Earle Buckingham, McGraw-Hill, New York, 1949.

$$E = \frac{0.86603 \times 0.57804}{0.86603 \times 0.57804 + 0.06} = .89\ (89\%)$$

$$\text{Power loss} = 100 - 89 = 11\%$$

As you see, the power loss is reduced substantially as the lead angle increases. This is true up to a lead angle of about 30°, after which there is little change in power loss.

12

Estimating Gear Size and Selection of Enclosed Gear Drives

In this chapter we use an example to show one procedure that could be followed to select a set of spur gears for a given, simple, application. This is followed by a discussion of the complexities that occur as the drive becomes more complex.

Finally we discuss the procedure to follow, again using an example, in selecting a speed reducer for a given application.

We start by listing the terms used in this chapter, alphabetically, some of which have been listed previously.

AGMA Q8

This denotes an American Gear Manufacturers Association Quality level, as determined from AGMA Standard 390.03. The tolerances for Q3 through Q15, tabulated in the appendix, pp. 164 through 170, are reprinted from this standard.

Allowable Compressive Stress (s_{ac})

This is the maximum allowable stress in compression for a given material, expressed in psi (pounds per square inch).

Brinell Hardness Number (BHN)

This denotes the hardness of a material as checked on a Brinell hardness checking machine.

Compressive Stress (s_c)

This is the stress that results when a load is pressed against a substance. It is expressed in psi (pounds per square inch).

Diametral Pitch (DP)

The number of teeth per inch of pitch diameter.

Face (F)

The width of that portion of a gear on which the teeth are cut. On a spur gear this is the length of the teeth.

Hertz, H (1857–1894)

A German physicist who derived the formula for the compressive stress that results when two metal cylinders are pressed together and the deformed area measured. His formula for compressive stress (s_c) is given on p. 101.

Horsepower (HP)

This term was first used by James Watt (1736–1819), a Scot engineer and inventor, credited with the development of the steam engine, one of several inventions that triggered the Industrial Revolution. To give prospective buyers of his engine an idea of its capabilities, he related its power to that exerted by a horse pulling a load.

It is necessary here to provide a couple of definitions. In Physics, there is a difference between the terms "work" and "power."

The term "work" means the overcoming of resistance, and the amount of work is measured in foot-pounds. Thus, a man exerting a force of 200 lbs to pull a cart for 10 ft does 2000 ft-lbs of work. This is so regardless of how long it takes to pull the cart for 10 ft.

"Power" is the amount of work done in a period of time. Thus if, in the above example, the man pulls the cart 10 ft in 1 min, he will have exerted 2000 ft-lbs per minute of power.

Watt calculated that a draft horse could exert 32,400 ft-lbs per minute, a figure later rounded to and standardized at 33,000, the standard to this day. Thus, by definition:

$$1 \text{ HP} = 33{,}000 \text{ ft-lbs/min (or 550 ft-lbs/sec)}$$

K Factor (K)

This is an index of the intensity of tooth load as it relates to surface durability.

Lewis, Wilfred

The developer of the "Lewis Formula," which has been used for years to determine the approximate bending stress at the root of a gear tooth. It was derived in 1893 from the formula for stress at the base of a cantilever beam. This formula is given on p. 102.

Modulus of Elasticity (E)

This is the ratio of stress over strain as determined from test bars. For steel, $E = 30{,}000{,}000$.

Number of Teeth (N_p, pinion; N_G, gear)

The number of teeth contained in the whole circumference of the pitch circle.

Pi (π)

See Appendix, p. 192.

Pitch Diameter of Pinion (d)

This is the diameter of the pitch circle of the pinion.

Pounds per Square Inch (psi)

A unit of measurement of an applied load.

Pressure Angle (ϕ)

The angle through which the pressure, or power, is transmitted from the driving to the driven gear teeth. See Figure 2.1b, "Construction of Involute Curves" in Chapter 2.

Ratio (m_g)

Same as "gear ratio," or "speed ratio."

Revolutions per Minute (rpm) or (N)

A unit of measurement of the speed of a rotating body.

Tangential Tooth Load (W_t)

The load impressed by the teeth of the driver to the teeth of the driven gear in a direction tangential to the pitch circles.

Tensile Stress (s_t)

This is the stress resulting from a load trying to stretch a bar; or in bending a bar, the side opposite the direction of the load is in tension.

Tooth Form Factor (Y)

Called the "Y factor." This was developed by Wilfred Lewis from the layout of a gear tooth to convert it to a simple cantilever beam. It is a dimensionless ratio, is tabulated, and can be used for any pitch.

Torque (T)

The load multiplied by its distance from the center of rotation. Thus the torque resulting from applying a tangential load on a 6″ *PD* gear tooth is $W_t \times 3.000$ in-lbs.

On pp. 100 through 105 we give an example of one way to select the size of a gearset to handle the assumed loads, speeds, etc. indicated.

To estimate spur pinion and gear diameters, center distance, pitch, and face width, follow the steps in this example, which is an open gearset driven by an electric motor through a speed reducer (the desired overall ratio is too high for a single gearset).

We start by making the following assumptions:

1. Medium accuracy (AGMA Q8)
2. Pinion hardness 225 BHN, gear hardness 210 BHN, 20° PA, not modified
3. 20 HP motor, 1750 rpm, coupled to input shaft of a 20:1 ratio speed reducer, with pinion mounted on reducer output shaft
4. Desired speed of gear shaft: 30 rpm (approx.)
5. Efficiency of reducer: 90%.

Then: Pinion speed = 1750/20 = 87.5 rpm.
Ratio = 87.5/30 = 2.917:1 (use 3:1).
Actual speed of gear shaft = 87.5/3 = 29.17 rpm.

To determine tangential tooth load (W_t): The relationship between horsepower and torque is expressed in this formula:

$$\text{HP} = \frac{2 \times \pi \times N \times T}{33{,}000}, \qquad \text{or} \qquad T = \frac{33{,}000 \times \text{HP}}{2 \times \pi \times N}$$

where

N = rpm of motor.
T = torque = pinion pitch radius ($d/2$) × W_t.
33,000 is ft-lbs per minute (× 12 = 396,000 inch-lbs per minute), which by definition = 1 HP.

Substituting assumed and known values in the equation for torque:

$$T = \frac{396{,}000 \times 20}{2 \times \pi \times 1750} \times 20 \times 0.90 = 12{,}965 \text{ inch-lbs}$$

$$W_t = \frac{12{,}965}{d/2} = \frac{25{,}930}{d} \text{ lb}$$

Since gears in the hardness range in this example are more apt to fail from surface stresses than tooth breakage, we will work from Hertz stresses, the formulas for which are derived from the work of H. Hertz of Germany. He used two steel cylinders, pressed them together, and measured the width of the deformed area. From his formula the formula for compressive stress is derived and is as follows:

$$s_c = \sqrt{\frac{0.70}{2/E \cos\phi \sin\phi}} \sqrt{\frac{W_t}{Fd} \frac{m_g + 1}{m_g}}$$

where

s_c = compressive stress
E = modulus of elasticity, which is stress/strain as determined from test bars. For steel, E = 30,000,000.
ϕ = pressure angle = 20° in this example
W_t = tangential tooth load
F = face width in inches
d = pitch diameter of pinion in inches
m_g = speed ratio = 3:1 in this example.

So for our example:

$$s_c = \sqrt{\frac{0.70}{2/30{,}000{,}000 \times .93969 \times .34202}} \sqrt{\frac{W_t}{Fd}\left(\frac{m_g + 1}{m_g}\right)}$$

$$= 5715 \sqrt{\frac{W_t}{Fd}\left(\frac{m_g + 1}{m_g}\right)}$$

It is convenient to call the term under the square root the "K factor," so the above becomes:

$$s_c = 5715 \sqrt{K}$$

In our example:

$$K = \frac{25{,}930/d}{Fd}\left(\frac{3 + 1}{3}\right) = \frac{25{,}930}{Fd^2} \times 1.333 = \frac{34{,}565}{Fd^2}$$

$$s_c = 5715 \sqrt{\frac{34{,}565}{Fd^2}} \qquad s_c^2 = \frac{32{,}661{,}225 \times 34{,}565}{Fd^2}$$

$$Fd^2 = \frac{32{,}661{,}225 \times 34{,}565}{s_c^2}$$

s_{ac} =allowable compressive stress = 100,000 psi (approx.) for steel in the assumed hardness range

$$Fd^2 = \frac{32,661,225 \times 34,565}{(100,000)^2} = 112.9$$

If $F = 3$, $d^2 = 37.63$, $d = 6.134$.
If $F = 2.5$, $d^2 = 45.16$, $d = 6.720$.

A 20° PA nonmodified pinion must have at least 18 teeth to avoid undercut, so it looks like the pitch we are talking about is 3 *DP*. We choose a 3″ face in case we need the added tooth bending strength over what it would be with a 2.5″ face.

Therefore

$$\text{No. of teeth in pinion } (N_p) = 6.134 \times 3.000 = 18.4 \text{ (use 19)}$$

$$\text{No. of teeth in gear } (N_g) = 19 \times 3 = 57$$

$$\text{Pinion } PD = 19/3.000 = 6{,}333'' \text{ and gear PD} = 57/3.000 = 19.000''$$

$$CD = (6.333 + 19.000)/2 = 12.667''$$

$$W_t = \frac{T}{d/2} = \frac{12,965 \times 2}{d} = \frac{25,930}{6.333} = 4094 \text{ lb}$$

To determine the approximate bending stress at the root of the teeth we will use the original Lewis formula, derived by Wilfred Lewis in 1893 from the formula for stress at the base of a cantilever beam loaded at the end. The Lewis formula is:

$$s_t = \frac{W_t \times DP}{F \times Y}$$

where

s_t = tensile stress
W_t = tangential tooth load
DP = diametral pitch
F = face width in inches
Y = tooth form factor, or "Y factor" = .287 (tip loading, from tables)

$$s_t = \frac{4094 \times 3.000}{3 \times .287} = 14,265 \text{ psi.}$$

The yield point of steel in the 200 BHN range is about 60,000 psi. Gear teeth have a root fillet stress concentration factor of close to 2 (Lewis did not use a stress concentration factor in his formula, but he allowed for it by using low design stresses). We will divide 60,000 by 2 for stress concentration and then by 2 for a margin of safety. This gives an approximate "Lewis" stress of 15,000 psi, which says that we are in the ball park on allowable bending stress.

There are, however, many more factors to consider before a final design is reached. Some of these are:

1. The characteristics of the driven machine. Is it apt to stall on occasion? If so the stall torque of the motor must be considered in calculating the breaking strength of the teeth. Does it vibrate? If so the dynamic load on the gear teeth will be increased.
2. The foregoing calculations assume full contact across the face. This almost never happens because of shaft misalignment, shaft deflection under load, helix angle error, etc.
3. Allowance must be made for the low pitch line velocity in this example (about 145 ft/min). This results in a marginal oil film thickness between meshing teeth, even when a relatively heavy oil is used. This is what happens with plain bearings—at low speeds the oil film is not properly distributed, but is improved as the speed increases.

All this requires that the gears be derated.

4. The number of hours per day the machine is expected to operate. Let's say that this is 8 hours per day, 5 days per week. The rpm of the pinion is 87.5; so the tooth contacts per minute of any one tooth are therefore 87.5. Tooth contacts per week = $87.5 \times 60 \times 40 = 210{,}000$, and for a 48-week year, 10,080,000 cycles per year.

The life of such a gearset, properly designed and assembled, is usually at least 10^7 cycles. This means that this gearset could probably last for 1 year even though it is operated almost continuously at full power. In most applications, though, gearsets will not operate much of the time at full power. In fact, they are generally not operated continuously, as in this case. On the basis of casual operation and limited usage at full power, this gearset would probably last for 5 years without undue pitting or wear. If the break-in conditions were favorable to polish the tooth surfaces, and if the kind of steel used would tend to work harden, a gearset like this might even last for 10 years under operating conditions of intermittent running and a relatively small percentage of time at full power.

The trick will be to complete the design to achieve a life of about 10 years, now following all the steps and gear trade criteria outlined in Darle Dudley's *Handbook of Practical Gear Design,* Chapter 3, pp. 3.9 to 3.153 and in applicable AGMA standards. It is suggested, however, that only experienced gear design engineers be entrusted with developing the final specifications of the gears.

The various changes in specifications that can be made to increase capacity include modification to increase recess action, increase hardness of both gear and pinion, improve accuracy, make certain care in assembly is adequate, etc.

Finally, if this gearset is to become a production item, then a few sample sets should be made up to the final specifications, assembled, and operated for a long enough time to determine if the gears are, in fact, properly designed.

The AGMA system for determining compressive and bending stresses modifies the basic formulas to provide for some of the derating requirements listed in p. 103. This is accomplished by use of factors that affect the loading on gear teeth.

First, the compressive stress and bending stress formulas are modified by geometric factors I and J, from ANSI/AGMA Standard 2001-B88, which is a revision of AGMA 218.01, as follows:

I is Pitting Resistance Geometry Factor. It is a dimensionless number, and takes into account the effects of the radii of curvature of the tooth profiles, the load sharing of meshing teeth, and the normal component of transmitted load. It is defined as:

$$\mathrm{I} = \frac{C_c\, C_x\, C_u}{m_N}$$

where

C_c = curvature factor at pitch line
C_x = contact height factor
C_u = helical overlap factor
m_N = load sharing ratio.

For spur gears the helical overlap factor of course does not apply. The value of I is inserted in the basic formula for s_c as a divisor, under the square root sign.

J is the Bending Strength Geometry Factor. It is also a dimensionless number, and takes into account the shape of the tooth, the worst load position, stress concentration, and load sharing between oblique lines of contact in helical gears. Both tangential and radial components of the tooth load are included. The J factor calculation must be repeated for both the pinion and the gear. The factor is defined as:

$$J = \frac{YC_u}{K_f m_N}$$

where

Y = tooth form factor
C_u = helical overlap factor
K_f = stress correction factor
m_N = load sharing ratio.

Here again the helical overlap factor of course does not apply for spur gears.

The value of J is substituted for Y in the Lewis formula, which becomes:

$$s_t = \frac{W_t \times DP}{F \times J}$$

Other derating factors are applied to these basic formulas. These are not given here, as this is a book on gear fundamentals, and as such is not intended to cover more than the basic steps in gear design. The AGMA Standards and the reference books listed in the appendix will provide all the information necessary for the aspiring gear designer.

The foregoing example is a very elementary one. Even it, however, becomes much more complex when the derating requirements listed on p. 103 are considered.

A recent AGMA publication, AGMA 901-A92 Information Sheet, would be helpful to anyone wishing to pursue derating procedures. The title of this information sheet is "A Rational Procedure for the Design of Minimum Volume Gears" (the word "volume" refers to size, not quantity). In addition, a more detailed procedure can be found in Darle W. Dudley's *Handbook of Practical Gear Design.*

An example of a much more complex problem would be the selection of gearing for the drive on a diesel cargo ship. The gears could be spur or helical (helicals are more common for ship drives, but spurs are in use), the speed reduction single or double stage, and the rating method one of several (AGMA, Lloyd's Registry, American Bureau of Shipping, to name a few).

In addition, the tooth form might be noninvolute. On some Russian-built ships the *Novikof* form has been used with good results.

It is clear that gear selection and design can be frought with danger for the relatively inexperienced designer. Certainly for the more complex applications it would be wise to seek out gear consultants who have had experience in the type of drive being considered.

For helical and bevel gears the same basic approach in their selection can be followed, adjusting for equivalent face width, pressure angle, and direction of loads.

To select a speed reducer for a given application, let us take the example given for the 20:1 ratio reducer on p. 100.

Ratio:	20:1
Drive:	20 HP, 1750 rpm electric motor
Nature of load:	Assume it is a uniformly loaded conveyor (application factor is 1.25, from table 12.1)

Table 12.1. Typical Application Factors

Application factor	Driven equipment
1.25	Uniformly loaded conveyors, pure liquid mixers, Centrifugal compressors, rotary pumps
1.50	Nonuniformly fed conveyors, variable density mixers, lobe compressors, reciprocating pumps
1.75	Multicylinder reciprocating compressors, rubber extruders
2.00	Reciprocating conveyors, single-cylinder compressors, laundry washers (commercial)

The reducer rating must therefore be $20 \times 1.25 = 25$ HP at 1750 rpm input.

Reference to a reducer manufacturer's catalog will provide the information necessary to select a unit.

13

Enclosed Gear Drives

Enclosed gear drives are called *speed reducers* (or speed increasers) when the enclosure, called the *housing,* contains the bearings, seals, shafts, etc., as well as the gears, as opposed to enclosures placed over open gearing for safety reasons or protection from dust. Speed reducer housings are usually cast iron for standard units, and steel weldments for low-quantity special units.

Most of the standard units listed by manufacturers of such drives are speed reducers, not increasers, although some of those listed could be used as speed increasers IF none of its pinions are modified oversize, or if they are, their mating gears are oversize by the same amount so that angles of approach and recess are approximately equal, as discussed on p. 30.

The Most Common Types of Speed Reducers are:

Parallel Shaft

These are single reduction (one gearset), double reduction (two gearsets), and triple reduction (three gearsets). The input shaft is driven by a motor through a V-belt drive or is direct connected through a coupling. The output shaft has either a spur pinion cut integral with it, or pressed on and keyed or splined, driving a large gear pressed on and keyed or splined to the shaft of the machine being driven, or is connected to the shaft to be driven by one of several types of couplings. The gears in these units are either helical or herringbone. In triple reduction units the low-speed gear and pinion are sometimes spurs, with the gear teeth induction hardened and the pinion carburized and hardened.

Shaft Mounted, Parallel Shaft

These are single or double reduction with input shaft and no output shaft. In this type of reducer the low-speed gear bearings are mounted on the outside diameter of its hubs and its bore slides onto the shaft to be driven, driving by means of a key or splines, often held in place by a Dodge (or equivalent) Taper-Lock bushing. Shaft-mounted reducers must be restrained from rotating. This is accomplished by connecting a rod, called the *torque arm,* to the reducer and a rigid part of the machine being driven, or by flange mounting. The gearing in these units is helical.

In-Line Reducers

These drives are double reduction, with the gearing arranged so that the input and output shafts are in line with each other. A common use for this type of drive is in a *gearmotor,* in which an electric motor is connected to the reducer by flange mounting. The gearing in these units is helical in integral horsepower drives. In fractional horsepower in-line drives the gearing is usually spur, fine pitch, narrow face.

Right-Angle Drives

There are three basic types. These are: worm gear reducers, comprised usually of one worm driving one worm gear, or one worm driving two worm gears if two output shafts are required; combination worm gear and helical reducers, in which the initial reduction is usually the worm and worm gear; and combination spiral bevel and helical reducers, in which the initial reduction is usually the spiral bevel pinion and gear.

All of these reducer types are catalogued by several gear and reducer manufacturers. To obtain their names and addresses write to American

Gear Manufacturers Association, 1500 King St., Alexandria, VA 22314, or call 703-684-0211.

Catalogs containing illustrations, dimensions, load ratings, etc. can be obtained from the reducer manufacturers or their distributers.

AGMA can provide the names of unmounted gear manufacturers as well.

Other Enclosed Drives

There are many other enclosed gear arrangements. A few of the most common are described in the following paragraphs.

Eplicyclic Gearing

This is a family of gear arrangements that includes two of the most common: *planetary* parallel shaft drives and the bevel gear *differential.*

The planetary parallel shaft drive is comprised of the *sun gear* in the center, meshing with three or more *planet gears* which also mesh with an internal gear, called the *annulus,* or *cage.* The planets must be equally spaced (at 120° when there are three). Depending on which member is fixed, which is input and which is output, the terms used to describe each are "planetary," "star," and "solar". These are illustrated in Figure 13.1a through 13.1c.

Table 13.1 provides data for each arrangement.

Planetary drive applications include automatic-shift automobile transmissions, aircraft propeller drives, and many others.

The *bevel gear differential's* most common application is in the rear axle drive of automobiles. Figure 13.2 is a schematic drawing of such a drive.

In the illustration, *S* is the drive shaft; *D* is the hypoid pinion; *E* the hypoid gear (bore not keyed); and *H* & *K* are the bevel side gears which are keyed to the wheel axles, one to the left wheel, the other to the right wheel. *T* are the studs that provide the bearings for the pinions *R* and by which the power from gear *E* is transmitted to the differential assembly as seen in the illustration. There are four differential pinions, spaced 90° apart.

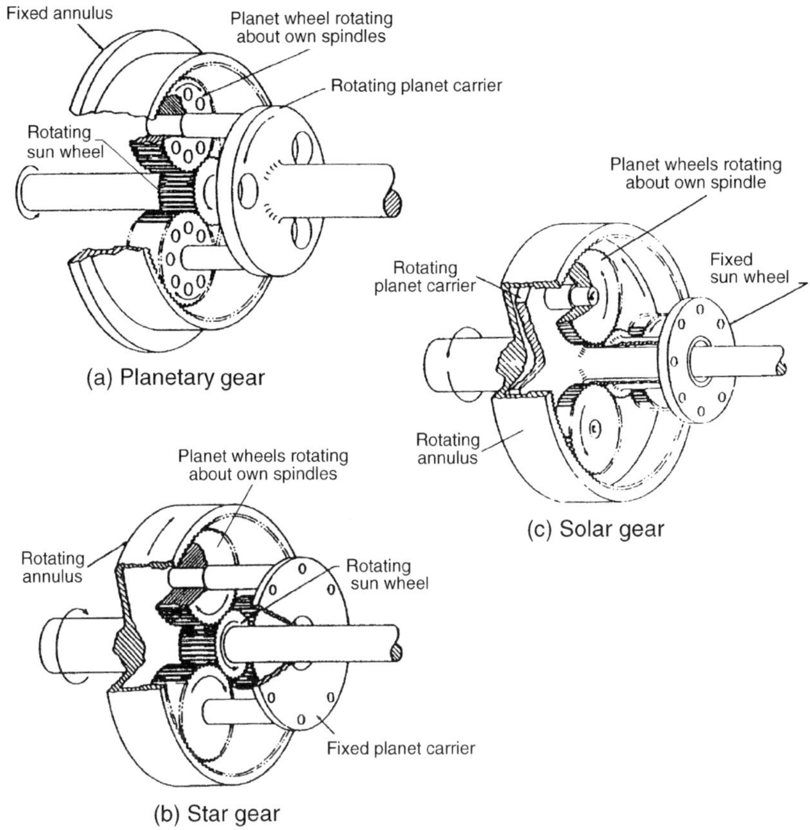

Figure 13.1. Simple Epicyclic Gears. (a) Planetary; (b) star gear; (c) solar gear. (Reprinted from *Gear Handbook,* Darle W. Dudley, Editor, McGraw-Hill, New York, 1962, with permission.) Note: See Table 13.1, p. 113 for epicyclic gear data.

When the car is going straight ahead *D* drives *E* and the differential gears revolve as a unit, with no gear tooth action between the side gears and the four pinions. Whenever the car starts to turn a corner, say to the right, the left wheel will have to travel further, so the left axle *B* must turn faster than *C*. Then the differential gears begin to move relative to each other, the action being that of an epicyclic gear train.

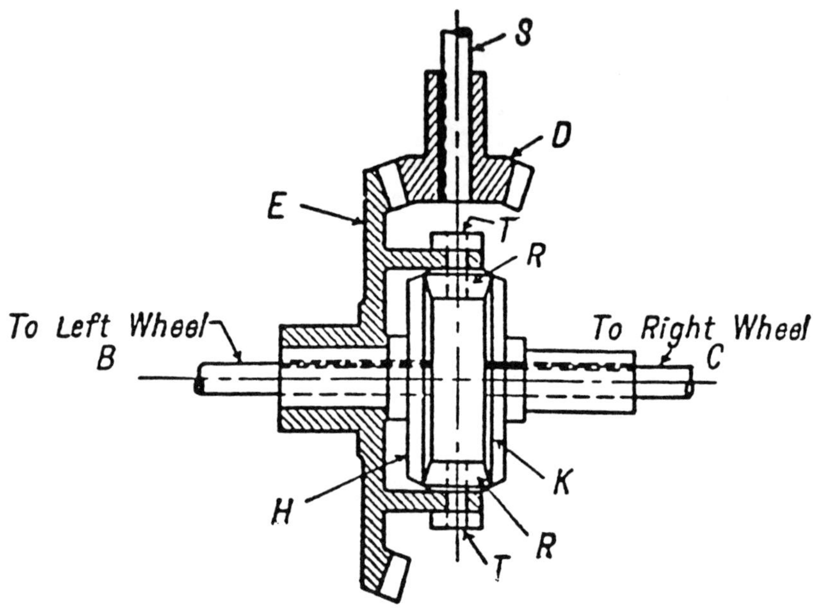

Figure 13.2. Bevel Gear Differential

Table 13.1. Single (Simple) Epicyclic Gear Data

Arrangement	Fixed member	Input member	Output member	Overall ratio	Range of ratios normally used
Planetary	Ring	Sun	Cage	N_r/N_s to 1	3:1 to 12:1
Star	Cage	Sun	Ring	N_r/N_s	2:1 to 11:1
Solar	Sun	Ring	Cage	$N_s/N_r + 1$	1.2:1 to 1.7:1

N_s = number of sun teeth

N_p = number of planet teeth

N_r = number of ring (annulus) teeth

To assemble: $N_r = N_s + 2N_p$

$\frac{N_r + N_s}{\text{No. of planets}}$ must equal a whole number.

Transmission Gearing

Probably one of the most widespread uses of gearing is in automotive, truck, off-road vehicles, etc. transmissions. These gears are helical, in *constant mesh,* the shifting from low to higher speeds and reverse being accomplished by toothed clutches sliding between each gearset. Illustrations of such transmissions are shown in Figures 13.3 and 13.4. These are manual transmissions. Figure 13.5 shows an automatic transmission.

An interesting aside is a story told to me by an old friend and mentor, the late Professor Earle Buckingham of MIT. When Ford Motor Company built their first Model "A" gear shift transmission it was noisy. Ford called in Prof. Buckingham to solve the problem. In those days (about 1928), improvement in gear accuracy could have helped but Ford had achieved about the most accurate hardened gearing possible at that time.

Prof. Buckingham spent a couple of days at Ford and in a few weeks sent Ford his recommendations. He had changed the number of teeth in each gear so that the mesh in each gearset was a harmonic with others in the transmission. The noise level did not change but the sound was musical, and that ended the complaints.

At about the time of the above solution for Ford's problem, Buckingham had been writing a book, *Spur Gears,* published in 1928. In Chapter

Figure 13.3. Five-speed manual transmission. (Reprinted from *Gear Handbook,* Darle W. Dudley, Editor, McGraw-Hill, New York, 1962, with permission. Photo courtesy of Dana Corporation, Toledo, OH.)

VI of that book there is a section entitled "Music of the Gears" in which he uses an automobile transmission as an example, stating that: "By keeping the numbers of teeth in these gears to harmonious ratios, we can avoid certain unpleasant sounds." I have been told that in addition to solving Ford's problem by this approach he solved the same problem for several other automobile manufacturers as well.

There are many more types and sizes of enclosed gear drives, too numerous to cover in this chapter, the purpose of which is to give you the basics and tell you where to look for more detailed information.

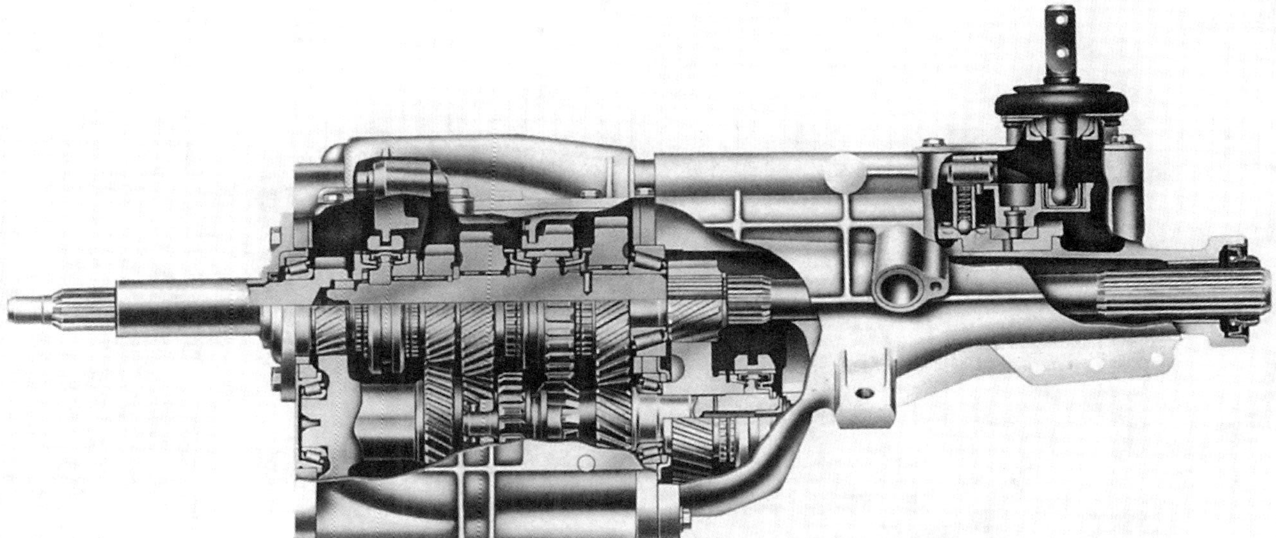

Figure 13.4. T-5 five-speed manual transmission. (Courtesy of Borg Warner Automotive Corporation, Muncie, IN.)

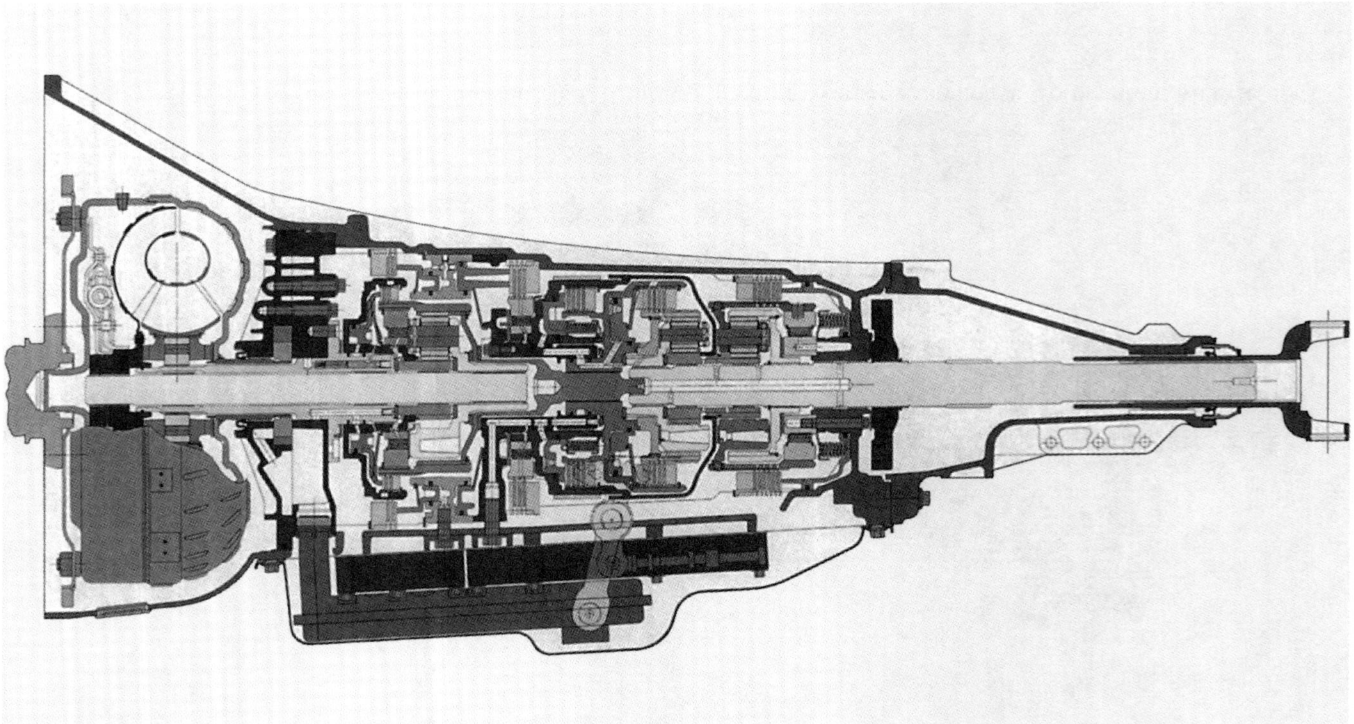

Figure 13.5. E40D automatic transmission. See Figures 13.6 to 13.8 for Element Applications Chart, Solenoid Logic, and Schematic Arrangement for this transmission. (Courtesy of Ford Motor Company, Dearborn, MI.)

GEAR RATIOS	MODE	COAST CL1	INTERMED CL2	DIRECT CL3	FORWARD CL4	REVERSE CL5	CONVERTER CL6	OVERDRIVE CL7	BAND B	ONE WAY CLUTCHES DRIVE OWC 1	DRIVE OWC 2	DRIVE OWC 3	COAST OWC 1	COAST OWC 2	COAST OWC 3
$\frac{S2}{R2}\left(\frac{R3}{S3}+1\right)+1=2.710$	1M	X			X	X									
	1A				X		•			X		X	O R		O R
$1+\frac{S2}{R2}=1.538$	2M	X	X		X		•		X			O R			O R
	2A		X		X		•			X	X	O R	O R	O R	O R
1.000	3M	X	X	X	X		•				O R	O R		O R	O R
	3A		X	X	X		•			X	O R	O R	O R	O R	O R
R1 R1 + S1 = .712	4		X	X	X		•	X		O R	O R	O R	O R	O R	O R
R3 S3 = 2.176	R	X		X		X					O R			O R	

• = Apply release based on EEC strategy X = Engaged O R = Overrunning
Gear Teeth: R1 = 94. P1 = 28. S1 = 38: R2 = 78. P2 = 18. S2 = 42: R3 = 74. P3 = 20. S3 = 34.

Figure 13.6. Ford E40D automatic transmission Element Applications Chart.

PANOL POSITION	GEAR	SHIFT CONTROL SOL. 1	SHIFT CONTROL SOL. 2	CONVERTER CLUTCH CONTROL SOL. 3	COAST CLUTCH CONTROL SOL. 4	COAST BRAKING
OD	4 3 2 1	OFF ↓ ON ↓	OFF ON ↓ OFF	ON OFF BASED ON EEC STRATEGY ↓	OFF ↓	YES NO ↓
OD (CANCEL) SW. ON	3 2 1	OFF ON ↓	ON ↓ OFF		ON ↓	YES NO ↓
2	2	OFF	OFF		OFF	YES
1	I	ON	OFF	OFF	OFF	YES

Figure 13.7. Ford E40D automatic transmission Solenoid Logic.

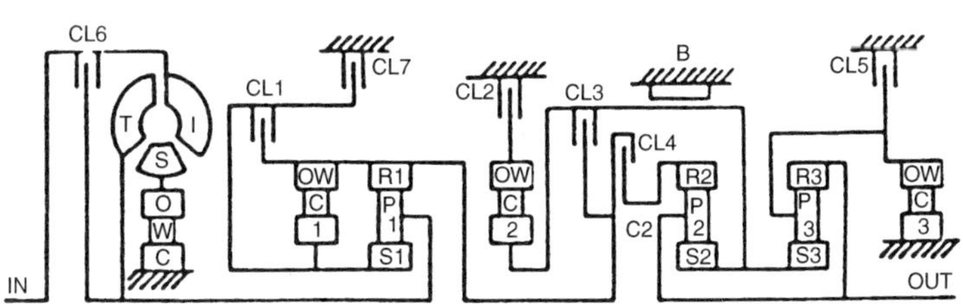

Figure 13.8. Ford E40D automatic transmission, Schematic Arrangement.

14

Noise and Vibration

A problem most gears have is that they make noise. Some types of gears make more noise than others but they all are guilty to some degree.

Comparing the various types of gears, assuming all are made to the same degree of accuracy and are operating at the same pitch line velocity, we list them in order of inherent noise level, listing the noisiest first:

Spur
Straight tooth bevel
Zerol®
Hypoid
Spiral bevel
Herringbone (continuous tooth)
Helical and double-helical
Worm gearing and Spiroid® gearing.

It is assumed further that in the above listing pinions and gears are both steel, except in worm gearing the gears are bronze. One way to quiet down a gearset is to make the pinion or gear from one of the nonmetallic materials, but this method has limited application.

The main causes of gear noise start with the design of the gearing. For example, spurs, helicals, and worm gearing should be designed for as much recess action as possible (see Figure 3.3) and spur and helical pinions must not have undercut. The helix angle on helicals must be great enough to provide adequate tooth overlap. On bevel gearing the face width should be no more than one third the cone distance, and straight tooth bevels should be specified as Coniflex®.

Properly designed gears can of course still be noisy if they are not made right. The most common causes of gear noise are:

1. Involute error
2. Tooth spacing error
3. Runout
4. Inadequate backlash
5. Gear blank resonance
6. Shaft torsional and longitudinal resonance
7. Housing resonance
8. Driven machinery resonances
9. High horsepower drives, such as 1000 to 10,000 HP or more

The noise generating conditions listed as items 1 through 4 are amplified when any or all of the resonance conditions listed as items 5 through 8 exist. High horsepower drives, number 9 above, are apt to be noisier than lower HP drives, with all else equal.

The term "resonance" used in items 5 through 8 is a term describing a state when an elastic body vibrates at its natural frequency, which can be determined by striking the body with a hammer and measuring the frequency of the resulting sound.

When the frequency of gear teeth in mesh is the same, or close to, the natural frequency of any of the above, each will respond, amplifying the sound of the tooth contacts. This can be so even if the gears are very accurate and the conditions of items 1 through 4 are minimal. So one of the things to look for in efforts to solve noise problems is the natural frequency of each part of a gear drive system.

I had some experience in this regard back in 1938, when I was a student engineer with General Electric Company in Lynn, Mass. I was assigned a problem G.E. was having with some of their steam turbines—those used to drive auxiliary equipment in power plants and ship engine rooms. Since turbines operate at high speeds the output speed had to be reduced with reduction gearing. And when direct connected or close coupled to the turbine output shaft, vibration of the turbine buckets, or blades, occurred. This was causing fatigue failures. Illustrations of a second row turbine bucket and a typical fatigue failure are shown in Fig. 14.1. There were no failures in first row buckets, which are shorter, and therefore had a higher natural frequency. Much of my investigation, with G.E.'s permission, was used as my thesis for the Master of Science degree at MIT. I quote from the introduction of my thesis as follows:

"Recently failures have occurred in the second row of blades, the longer of the two rows, in steam turbines of the type described above. These are fatigue failures which occur at the dovetail joint of the blade and progress in an axial direction, the direction of the progressive

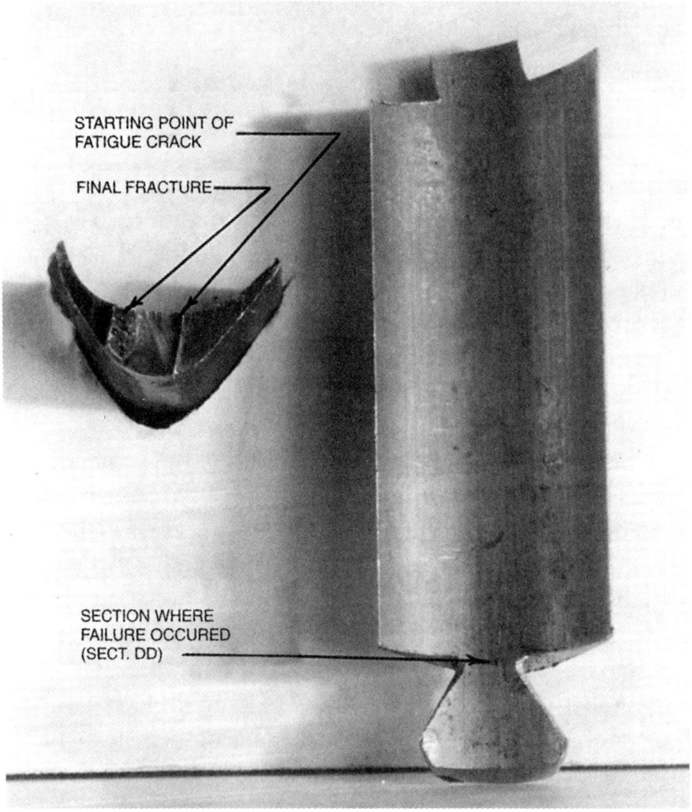

Figure 14.1. Second row turbine buckets showing service fatigue failure. (From *Axial Vibration of Steam Turbine Buckets.* Master's thesis by Richard H. Ewert, MIT, 1938.)

failure being easily determined by referring to the photograph, on page 120, of a blade which actually failed in service.

> "It is well known that this type of failure is the result of repetition of stress, the crack usually starting at some point of high stress concentration. We see immediately from the geometry of the bucket that the failure occurred at the point where the stress concentration is highest. . . . And we conclude that the repetition of stress results from the blade vibrating . . . at some one of its natural periods. This fact leads us to the further conclusion that some part of the turbine is exciting this vibration with impulses which are in resonance with one of the natural frequencies of the blades, and probably acting in an axial direction.
>
> "From a study of the service records of the turbines of this type, it is seen that where a flexible coupling was used to connect the turbine shaft to the reduction gears, or where no gears were used at all, no blade failures occurred. But in all of the five turbines where blades failed, the turbine shaft was direct connected, or close coupled, to reduction gears. In these the pinion of the gears is either shrunk on the shaft, or in some cases cut in the shaft itself."*

The foregoing passage from my thesis suggests that the gears were causing the vibration. But were there other contributing factors? It turned out that there were, among them resonance conditions in various components which amplified the vibration energy as it traveled from gear teeth to turbine buckets.

It was found that the natural frequency of the turbine wheel web and the longitudinal natural frequency of the turbine wheel shaft were close enough to the gear tooth contact frequency at normal operating speed to cause trouble. A redesign could have perhaps solved the problem, but a simpler and less costly solution presented itself—provide a means to dampen the vibration where it was causing trouble.

It was found that some of the blades did not vibrate at all if they were relatively loosely held in the wheel, telling us that friction at the joint was dampening the vibration. One of my recommendations was

---

*From *Axial Vibration of Steam Turbine Buckets,* by Richard H. Ewert, Thesis submitted as partial requirement for BS and MS degrees at MIT, 1938.)

therefore to change the tolerances of the bucket dovetail so that all would be dampened, or to change the configuration of the dovetail by crowning it in the axial direction, permitting it to move slightly instead of vibrating.

There are other ways to combat gear noise, one of which is suggested by Earle Buckingham's solution for Ford Motor Co. back in 1928, referred to in the preceding chapter, p. 113.

In any gear drive with two or more reductions, the noise generated by gear teeth can be made less abrasive if the tooth contacts per second in each mesh produce sounds that are harmonious. Often a change of a few teeth in the gearing can accomplish this. With today's greater gear accuracy, as compared with what could be achieved in 1928, this is seldom considered today, but it is interesting to hear some of Prof. Buckingham's observations.

I made previous reference to his book *Spur Gears* in which there is a section he calls "Music of the Gears." In that section he gives us an elementary lesson in music theory, some of which I repeat here.

> "The extent to which the science of musical sounds may enter into machine design is but slightly appreciated. A successful engineer or mechanic has a certain instinct in this direction—he must have—but it does not appear that any definite connection is generally recognized. While the study both of dynamics and of sound are included under the same general heading—Physics—it seems, in practice, a far cry from the study of machine design, for instance, to the study of music. There is, however, a close connection.
>
> "The distinction between music and noise is, generally speaking, a distinction between the agreeable and disagreeable.
>
> "For commercial gear drives, where all noise cannot be eliminated, there is a demand for some degree of quietness; that is, what sound is present should be unnoticeable or agreeable. These less noticeable sounds are generally found in the lower pitches or tones. . . .
>
> "Noise in the operation of gears is caused primarily by errors in individual tooth action. The ear is capable of hearing sounds arising from 32 to 38,000 vibrations per second. Thus, when the number of tooth engagements per second exceeds 32, the result is a continuous noise, which may be recognized as a tone whose note or vibration period is designated by the number of tooth engagements made per second. In case these vibrations should exceed 38,000 per

second, no tone would be heard from this cause, but such speeds are beyond the range of ordinary gear practice. . . .

"The starting point of the musical scale, that is, the number of vibrations per second for the key note, is entirely a matter of convention. In practice there is a lot of disagreement. The ratios between the various tones, however, are constant, regardless of the exact number of vibrations per second selected as the starting point. The arbitrary scale of pitch in common use, and typified by the white keys of a piano, is represented by the following frequencies, which are very close to the actual number of vibrations per second ordinarily used:

Note	Frequency
c	128.0
d	144.0
e	160.0
f	170.6
g	192.0
a	213.3
b	240.0
c′ (middle C on piano)	256.0

"Thus, if a pair of gears revolved at such a speed that 256 tooth contacts were made each second, the pitch of the sound would be approximately that of middle C on the keyboard of a piano.

"*Resonance.* When we are dealing with a single pair of gears, this study is primarily one of resonance, the gears being the generator or source of the vibration, and the carrier or case containing the gears, the resonator. We have also to consider the natural tone frequency of the gears themselves. . . . If the case containing the gears has a natural tone frequency of, say, 256 vibrations per second, it will respond to the tone of the gears when their speed is such that the number of contacts made by the gear teeth is 256, or in that neighborhood. If the tone frequency of either of the gear blanks has this same natural frequency, this will also tend to enlarge upon the sound produced. Under these conditions, the case containing the gears becomes an amplifier, so that the result is a far greater volume of sound than originally started from the gear teeth themselves. It should be understood that the resonator can produce no noise which does not originate from some other source; in this example, the gears themselves. The study of violin construction resolves itself into the production of a resonator (violin body) that will respond equally to all tones. The ideal gear case would be one that would not respond to any tone.

"When two or more pairs of gears are engaged, we have to consider consonance in addition to the study of resonance. The combination of tones produced by the various gears may be discordant or harmonious, according to the ratio of their tone frequencies to each other. In general, the difference between consonance and dissonance, or between harmony and discord, or between music and noise, is that noise is unpleasant or irritating while music is heard either with pleasure or with indifference. . . .

"*Harmonious Ratios.* The general law governing consonance and dissonance, as stated by Dr. Helmholtz, in his "Sensations of Tone," is simplicity itself:

"Generally those tones, and only those tones, harmonize whose fundamental tones bear to one another ratios expressed by small numbers; and the smaller the numbers which express the ratios of vibrations, the more perfect is the harmony of the two sounds.

"For example, the following ratios give pleasing combinations of tones:

Ratio	Tone interval
1:1	Single tone
1:2	Octave
1:3	Octave and perfect fifth
1:4	Two octaves
1:5	Two octaves and major third
1:6	Two octaves and perfect fifth
2:3	Perfect fifth
2:5	Octave and major third
3:4	Perfect fourth
3:5	Major sixth
4:5	Major third
5:6	Minor third

except when the ratio gives an even octave, such as 1:8, three octaves, 1:16, four octaves, etc.; when the figure that expresses the exact ratio becomes larger than six, we begin to run into discords. For example, 8:9 is a major second and is more discordant than harmonious."†

I have quoted from Buckingham's *Spur Gears* more extensively than I had at first intended. One reason is that I felt the concept of "gear music" might be something that many of my readers have never heard of and would find interesting. Another is that I believe this has historic significance, particularly as regards the Ford transmission story.

It is important to note here that where there is noise there is vibration. And vibration can cause trouble, as illustrated by my turbine bucket failure. So it is not enough to make noise unobtrusive or to insulate it. Its source must be analyzed and every element of a drive system examined to determine if any element has a natural frequency at or near the source frequency. Once determined it is usually relatively easy to dampen it or to change it in some manner.

To provide more recent and similar views, I quote from *Dudley's Gear Handbook,* Dennis P. Townsend, Editor, Chapter 13, Gear Vibration by Darle W. Dudley:

> "Vibration in vehicles or air craft is unwelcome—even if the gears are quite able to run for a long time without failure. Vibrations induced in a closely connected mechanical system may cause premature failure of somewhat fragile components such as instruments, control mechanisms, small motors or even bolts. (Critical bolted joints may suffer from bolts coming loose.);
>
> and from Chapter 14, Gear Noise by Prof. Donald Houser of Ohio State University:
>
> "Gear noise is an often overlooked, yet very important gear design consideration. Due to increased emphasis on noise regulation and an increase in customer awareness of noise, the importance of gear noise in design has increased significantly in recent years. Where gears are used for power transmission, gear noise is of particular concern because it is often close to a pure tone, has a high sound pressure level, and its predominant frequencies are often near the ear's most sensitive region."‡

---

†From *Spur Gears* by Earle Buckingham, McGraw-Hill, New York, 1928.

‡From *Dudley's Gear Handbook,* Dennis P. Townsend, Editor, McGraw-Hill, New York, 1991. Reprinted with permission.

15

Lubrication

Improperly lubricated gears won't last very long. This is true, of course, for any machine part subjected to friction.

This chapter provides lubricant classification and application and servicing guidelines for industrial gearing which has been designed and rated by AGMA standards.

These guidelines apply to open gearing whose bearings are lubricated separately and to enclosed gearing when bearings and gears share a common lubricant.

Gear units or gearsets covered in this chapter are:

Helical

Herringbone (continuous tooth)

Double-helical

Spur

Bevel, straight tooth or spiral

Worm gearing (for enclosed drives only).

Types of Lubricants

The types of lubricants covered by ANSI/AGMA Standard 9005-094 are as follows:

Table 15.1. Viscosity Ranges for AGMA Lubricants

Rust and oxidation inhibited gear oils AGMA Lubricant No.	Viscosity ranges[a] mm^2/s (cSt) at 40°C	Equivalent ISO grade[a]	Extreme pressure gear lubricants[b] AGMA Lubricant No.	Synthetic gear oils[c] AGMA Lubricant No.
0	28.8 to 35.2	32		0S
1	41.4 to 50.6	46		1S
2	61.2 to 74.8	68	2EP	2S
3	90 to 110	100	3EP	3S
4	135 to 165	150	4EP	4S
5	198 to 242	220	5EP	5S
6	288 to 352	320	6EP	6S
7,7 Comp[d]	414 to 506	460	7EP	7S
8,8 Comp[d]	612 to 748	680	8EP	8S
8A Comp[d]	900 to 1100	1000	8AEP	—
9	1350 to 1650	1500	9EP	9S
10	2880 to 3520	—	10EP	10S
11	4140 to 5060	—	11EP	11S

continued

Table 15.1. continued

Rust and oxidation inhibited gear oils AGMA Lubricant No.	Viscosity ranges[a] mm^2/s (cSt) at 40°C	Equivalent ISO grade[a]	Extreme pressure gear lubricants[b] AGMA Lubricant No.	Synthetic gear oils[c] AGMA Lubricant No.
12	6120 to 7480	—	12EP	12S
13	190 to 220 cSt at 100°C (212°F)[e]	—	13EP	13S

Residual compounds[f] AGMA Lubricant No.	Viscosity ranges[e] cSt at 100°C (212°F)
14R	428.5 to 857.0
15R	857.0 to 1714.0

[a]Per ISO 3448, *Industrial Liquid Lubricants—ISO Viscosity Classification;* also ASTM D 2422 and British Standards Institution B.S. 4231.

[b]Extreme pressure lubricants should be used only when recommended by the gear manufacturer.

[c]Synthetic gear oils 9S to 13S are available but not yet in wide use.

[d]Oils marked Comp are compounded with 3% to 10% fatty or synthetic fatty oils.

[e]Viscosities of AGMA Lubricant Number 13 and above are specified at 100°C (210° F) as measurement of viscosities of these heavy lubricants at 40° C (100° F) would not be practical.

[f]Residual compounds—diluent type, commonly known as solvent cutbacks, are heavy oils containing a volatile, nonflammable diluent for ease of application. The diluent evaporates leaving a thick film of lubricant on the gear teeth. Viscosities listed are for the base compound without diluent.

CAUTION: These lubricants may require special handling and storage procedures. Diluent can be toxic or irritating to the skin. Do not use these lubricants without proper ventilation. Consult lubricant supplier's instructions.

Rust and oxidation inhibited gear oils ('R & O' gear oils). These are petroleum based and include chemical additives to resist rust and oxidation.

Compounded gear lubricants. These are a blend of petroleum base liquids with 3% to 10% of fatty or synthetic fatty oils. These lubricants are frequently used in worm gear drives.

Extreme pressure lubricants ("EP" lubricants). These are petroleum based liquids with chemical additives that produce a protective film to provide antiscuffing properties.

Synthetic gear lubricants. These are produced chemically to provide special properties to enhance performance or accommodate severe operating conditions. These lubricants have some disadvantages such as compatibility with other lube systems and behavior in the presence of moisture.

Viscosity ranges for the above lubricant types are given in Table 15.1.

Residual compounds are straight mineral oils in the heavier grades. They are normally mixed with a volatile diluent for ease of application. After the diluent evaporates the heavier material remains.

Special compounds. These and some greases are available. The viscosity of the base oil and the amount of solid lubricant, such as graphite, must be specified.

Lubricants for plastic gearing. Often a lubricant is an additive to the molding process, requiring no additional lubrication.

For lubricant number guidelines see tables 15.2 to 15.7.

Application Methods

There are several methods of application of lubricants. The most common are:

Table 15.2. AGMA Lubricant Number Guidelines for Enclosed Helical, Herringbone, Straight Bevel, Spiral Bevel, and Spur Gear Drives[a]

	AGMA lubricant numbers,[a,d,e] ambient temperature °C (°F)[f,g]			
Pitch line velocity[b,c] of final reduction stage	−40 to −10 (−40 to +14)	−10 to +10 (14 to 50)	10 to 35 (50 to 95)	35 to 55 (95 to 131)
Less than 5 m/s (1000 ft/min)[h]	3S	4	6	8
5 to 15 m/s (1000 to 3000 ft/min)	3S	3	5	7
15–25 m/s (3000 to 5000 ft/min)	2S	2	4	6
Above 25 m/s (5000 ft/min)[h]	0S	0	2	3

[a]AGMA lubricant numbers listed above refer to R&O and synthetic gear oil shown in Table 15.4. Physical and performance specifications are shown in Tables 15.1 and 15.3. EP or synthetic gear lubricants in the corresponding viscosity grades may be substituted where deemed acceptable by the gear drive manufacturer.

[b]Special considerations may be necessary at speeds above 40 meters per second (8000 feet per minute). Consult gear drive manufacturer for specific recommendations.

[c]Pitch line velocity replaces center distance as the gear drive parameter for lubricant selection. The corresponding table from the previous standard is included as annex B for reference.

[d]Variations in operating conditions such as surface roughness, temperature rise, loading, speed, etc., may necessitate use of a lubricant of one grade higher or lower. Contact gear drive manufacturer for specific recommendations.

[e]Drives incorporating wet clutches or overrunning clutches as backstopping devices should be referred to the gear manufacturer as certain types of lubricants may adversely affect clutch performance.

[f]For ambient temperatures outside the ranges shown, consult the gear manufacturer.

[g]Pour point of lubricant selected should be at least 5° C (9° F) lower than the expected minimum ambient starting temperature. If the ambient starting temperature approaches lubricant pour point, oil sump heaters may be required to facilitate starting and ensure proper lubrication.

[h]At the extreme upper and lower pitch line velocity ranges, special consideration should be given to all drive components, including bearing and seals, to ensure their proper performance.

Splash is the simplest method of applying lubricants to relatively low-speed gearing, in which the gear or an idler in mesh with the gear dips into the lubricant. Oils for this system are shown in Table 15.5.

Gravity feed or forced dip involves one or more oilers or a pan which allows oil to drip into the gear mesh at a set rate, again for low-

Table 15.3. AGMA Lubricant Number Guidelines for Enclosed Cylindrical Worm Gear Drives[a]

	AGMA lubricant numbers,[a] ambient temperature °C (°F)[c,d]			
Pitch line velocity[b] of final reduction stage	−40 to −10 (−40 to +14)	−10 to +10 (14 to 50)	10 to 35 (50 to 95)	35 to 55 (95 to 131)
Less than 2.25 m/s (450 ft/min)	5S	7 Comp	8 Comp	8S
Above 2.25 m/s (450 ft/min)	5S	7 Comp	7 Comp	7S

[a]AGMA lubricant numbers listed above refer to compounded R&O oils and synthetic oils shown in Table 15.4. Physical and performance specifications are shown in Tables 15.1 and 15.3. Wormgear drives may also operate satisfactorily using other types of oils. Such oils should be used, however, only with approval of the gear manufacturer.

[b]Pitch line velocity replaces center distance as the gear drive parameter for lubricant selection. The corresponding table from the previous standard is included as annex B for reference.

[c]Pour point of the oil used should be at least 5° C (9° F) lower than the minimum ambient temperature expected.

[d]Wormgear applications involving temperatures outside the limits shown above, or speeds exceeding 2400 rpm or 10 m/s (2000 ft/min) sliding velocity, should be referred to the manufacturer. In general, for higher speeds a pressurized lubrication system is required along with adjustments in recommended viscosity grade.

speed gearing. Recommended oils and rates of application are shown in Table 15.6.

Intermittent mechanical spray systems. Heavy oil is used which will stay on the gear teeth through several revolutions, activated automatically at timed intervals, usually once every revolution of the gear. See Table 15.6.

Continuous pressure lubrication. This method uses a circulating system with a pump, producing a continuous spray to the gear teeth. See Table 15.5.

Application by a brush, stick, or by hand. This is for the heaviest grades of lubricant, heavy pitch, and very-low-speed gears. An example of such a drive is the antenna final drive gear shown on p. 43. To determine how often application must be made close observation is required.

Table 15.4. AGMA Lubricant Number Guidelines for Enclosed Double-Enveloping Worm Gear Drives[a]

Center distance of final reduction stage	Worm speed of final reduction stage, rpm	AGMA Lubricant numbers[a] ambient temperature, °C (°F)[b,c]			
		−40 to −10 (−40 to +14)	−10 to +10 (14 to 50)	10 to 35 (50 to 95)	35 to 55 (95 to 131)
Up to 150 mm (6 inches)	Up to 700	7S	8 Comp	8A Comp	7S
	Over 700	7S	8 Comp	8 Comp	7S
Over 150 mm (8 inches)	Up to 450	7S	8 Comp	8A Comp	7S
To 300 mm (12 inches)	Over 450	7S	8 Comp	8 Comp	7S
Over 300 mm (12 inches)	Up to 300	7S	8 Comp	8A Comp	7S
To 450 mm (18 inches)	Over 300	7S	8 Comp	8 Comp	7S
	Up to 200	7S	8 Comp	8A Comp	7S
Over 450 mm (18 inches)	Over 200	7S	8 Comp	8 Comp	7S

[a]AGMA lubricant numbers listed above refer to compounded R&O oils and synthetic oils shown in Table 15.4. Physical and performance specifications are shown in Tables 15.1 and 15.3. EP oils with sulfur–phorphorus additives are not recommended for use without the gear manufacturer's approval. Wormgear drives may operate satisfactorily using other types of oils. Such oils should be used, however, only with approval of the gear manufacturer.

[b]Pour point of the oil used should be at least 5° C (9° F) lower than the minimum ambient temperature expected.

[c]Wormgear applications involving temperatures outside the limits shown above, or speeds exceeding 2400 rpm or 10 m/s (2000 ft/min) sliding velocity, should be referred to the manufacturer. In general, for higher speeds a pressurized lubrication system is required along with adjustments in recommended viscosity grade.

Table 15.5. *AGMA Lubricant Number Guidelines for Open Gearing (Continuous Method of Application)*[a,b]

Ambient temperature[c] in degrees Celsius (F)	Character of operation	Pressure lubrication Pitch line velocity: Under 5 m/s (1000 ft/min)	Pressure lubrication Pitch line velocity: Over 5 m/s (1000 ft/min)	Splash lubrication Pitch line velocity: Under 5 m/s (1000 ft/min)	Splash lubrication Pitch line velocity: 5 to 10 m/s (1000 to 3000 ft/min)	Idler immersion Pitch line velocity: Up to 1.5 m/s (300 ft/min)
−10 to 15[d] (15 to 60)	Continuous	5 or 5EP	4 or 4EP	5 or 5EP	4 or 4EP	8 to 9 8EP to 9 EP
	Reversing or frequent "start stop"	5 or 5EP	4 or 4EP	7 or 7EP	6 or 6EP	8 to 9 8EP to 9EP
10 to 50[d] (50 to 125)	Continuous	7 or 7EP	6 or 6EP	7 or 7EP	8 to 9[f] 8EP to 9EP	11 or 11EP
	Reversing or frequent "start stop"	7 or 7EP	6 or 6EP	9 to 10[e] 9EP to 10 EP		

[a]AGMA lubricant numbers listed above refer to gear lubricants shown in Table 15.4. Physical and performance specifications are shown in Table 15.1 and 15.2. Although both R & O and EP oils are listed, the EP is preferred. Synthetic oils in the corresponding viscosity grades may be substituted where deemed acceptable by the gear manufacturer.

[b]Does not apply to worm gearing.

[c]Temperature in vicinity of the operating gears.

[d]When ambient temperatures approach the lower end of the given range, lubrication systems must be equipped with suitable heating units for proper circulation of lubricant and prevention of channeling. Check with lubricant and pump suppliers.

[e]When ambient temperature remains between 30° C (90° F) and 50° C (125° F) at all times, use 10 or 10 EP.

[f]When ambient temperature remains between 30° C (90° F) and 50° C (125° F) at all times, use 9 or 9 EP.

Table 15.6. AGMA Lubricant Number Guidelines for Open Gearing Intermittent Applications[a–c] *[Where Gear Pitch Line Velocity Does Not Exceed 7.5 Meters per Second (1500 ft/min)]*

Ambient temperature[d] in degrees Celsius (F)	Intermittent spray systems[e]			Gravity feed or forced drip method[g]	
	R&O or EP lubricant	Synthetic lubricant	Residual compound[f]	R&O or EP lubricant	Synthetic lubricant
−10 to 15 (15 to 60)	11 or 11EP	11S	14R	11 or 11EP	11S
5 to 40 (40 to 100)	12 or 12EP	12S	15R	12 or 12EP	12S
20 to 50 (70 to 125)	13 or 13EP	13S	15R	13 or 13EP	13S

[a]AGMA Viscosity number guidelines listed above refer to gear oils shown in Table 15.4.

[b]Does not apply to wormgearing.

[c]Feeder must be capable of handling lubricant selected.

[d]Ambient temperature is temperature in vicinity of the gears.

[e]Special compounds and certain greases are sometimes used in mechanical spray systems to lubricate open gearing. Consult gear manufacturer and spray system manufacturer before proceeding.

[f]Diluents must be used to facilitate flow through applicators.

[g]EP oils are preferred, but may not be available in some grades.

Table 15.7. Lubricant Quantity Guidelines for Intermittent Methods of Application: Automatic, Semiautomatic, Hand Spray, Gravity Feed, or Forced Drip Systems [Where Gear Pitch Line Velocity Does Not Exceed 7.5 Meters per Second (1500 ft/min)]

Gear diameter	1/4 Hour[a] Face width in meters (inches)					1 Hour[a] Face width in meters (inches)					2 Hours[a,b] Face width in meters (inches)				
in meters (ft)	0.2 (8)	0.4 (16)	0.6 (24)	0.8 (32)	1.0 (40)	0.2 (8)	0.4 (16)	0.6 (24)	0.8 (32)	1.0 (40)	0.2 (8)	0.4 (16)	0.6 (24)	0.8 (32)	1.0 (40)
3.0 (10)	5.9	8.9	11.8	14.8	17.8	23.7	35.5	47.3	59.2	71.0	59.1	88.8	118	148	178
3.7 (12)	8.9	8.9	11.8	14.8	17.8	35.5	41.4	53.3	65.1	76.9	88.8	104	133	163	192
4.3 (14)	8.9	11.8	14.8	17.8	20.7	41.4	47.3	59.2	71.0	82.8	104	118	148	178	207
4.9 (16)	11.8	14.8	17.8	20.7	23.7	47.3	59.2	71.0	82.8	94.7	118	148	163	207	237
5.5 (18)	14.8	17.8	20.7	23.7	26.6	59.2	71.0	82.8	94.7	107	148	178	207	237	266
6.1 (20)	17.8	20.7	23.7	26.6	29.6	71.0	82.8	94.7	107	130	178	207	237	266	325
6.7 (22)	20.7	23.7	26.6	29.6	32.5	82.8	94.7	107	118	142	207	237	266	296	355
7.3 (24)	23.7	26.6	29.6	32.5	35.5	94.7	107	118	130	154	237	266	296	325	385
7.9 (26)	26.6	29.6	32.5	35.5	38.5	107	118	130	142	166	266	296	325	355	414
8.5 (28)	29.6	32.5	35.5	38.5	41.4	118	130	142	154	178	296	325	355	385	444

[a]The spraying time should equal the time for 1 and preferably 2 revolutions of the gear to insure complete coverage. Periodic inspections should be made to insure that sufficient lubricant is being applied to give proper protection.

[b]Two hours is the maximum interval permitted between applications of lubricant. More frequent application of smaller quantities is preferred. However, where diluents are used to thin lubricants for spraying, intervals must not be so short as to prevent diluent evaporation.

Tables 15.1 to 15.7 are reproduced from AN51/AGMA 9005-D94, with permission.

16

Costs

Selection of materials, heat treatments, gear tooth data, and tolerances all will affect costs. Attention to the purely geometric means to improve gear tooth action, discussed in Chapter 4, will help keep costs to a practical minimum. For example, it costs no more to hob or shape an oversize pinion than a standard one. Recess-action worm gearing is no more costly than conventional. Crowned teeth are little if any more expensive to produce than straight teeth.

Costs are affected most drastically, probably, by the tolerances specified. In this regard it is recommended that a painstaking study be made of the tables on pp. 164 through 170 in the appendix. These are reproduced from AGMA Standard 390.03. Tolerances for AGMA Q3 through Q15 are tabulated. There are later standards (ANSI/AGMA 2000-88 for spur and helical gears and AGMA 390.03A for bevel, hypoid, fine pitch worm gearing, and racks), but I have used 390.03 because I think these tables are easier to understand than the later tables, and the values are the same. Intelligent use of these tables still requires study of the AGMA publications indicated above.

These and other AGMA Standards that could be of interest to an aspiring gear designer are listed in the appendix on p. 162.

Now let's define "costs": The cost of a gear (or any machined part) is made up of the costs of material, subcontract costs, labor, and shop overhead (or "burden"). For most gears the major costs are labor and burden. To determine what these are it is necessary, first, to measure, and then record, the machine time for each machining operation, then convert the time into dollars. Determination of material and subcontract costs is easy. Determination of labor and burden costs is not easy, and must be done properly to avoid problems. It must be remembered that in a free market competitive environment, cost control can determine whether one gets or loses business, and if successful in getting, whether one keeps it.

We will start with a couple of definitions:

1. *Direct labor cost:* The wages paid to employees who work directly on the product, excluding fringe benefits (these, as well as indirect labor costs, are part of factory burden).
2. *Burden costs:* The total of all manufacturing, material handling, factory services and general factory expenses.

Allocation of Costs

It is no problem to accumulate total direct labor and factory burden costs over a period of time, totalling each at the close of each accounting period, and presenting them as part of the typical P & L statement.

It *is* a problem to allocate, equitably, direct labor and factory burden costs to individual job orders. This must be done in some manner before a job becomes an order, in estimating the cost of each job preparatory to submitting a price quotation. For control purposes, after a job becomes an order, it is important that factory costs be recorded in the same "language" used in the estimate. There are a number of ways to allocate labor and burden costs to cost estimates and individual job orders. Two of the most common are given in the following subsections.

Direct Labor Plus Shop Average Burden Percentage

A common procedure is to record direct labor times on job order sheets, convert times to dollars, then multiply the direct labor cost by a percentage to cover burden costs, having determined the percentage to use from historical data. This system possibly is satisfactory for a captive operation, where only overall costs are the primary concern of management. It has pitfalls, however, particularly in a shop engaged in jobbing work, as it ignores the fact that some machines cost more to operate, excluding the cost of direct labor, than others. This results in the reported "cost" on some machines being too high, others too low. Also there is the problem of reporting direct labor costs on multiple operated equipment as the machines-per-man ratio will vary from day to day in the average shop. This presents a problem in developing meaningful standards of performance as the reported cost on a job cannot be directly compared to the original estimate. It presents a further problem in estimating costs, as a machines-per-man ratio must be assumed in developing the estimate which could turn out to be too high a ratio when the job is run.

The Machine-Hour Rate System

The machine-hour rate method of allocating costs permits both the estimation and reporting of costs on the basis of machine time for each machining operation. It is particularly useful in job shop estimating and costing, where, as in the typical gear or general machine shop, investment in equipment is large and varies greatly from machine to machine, and where many machines are multiple operated (one man operates two or more machines).

With this system a rate is developed for each machine that represents each machine's proportionate share of direct labor and factory burden costs. Proportionate direct labor costs are determined from the top labor rate for each machine divided by its machines-per-man ratio. Proportionate factory burden costs are determined by first projecting such costs, then allocating them equitably to each machine, then dividing by the expected operating hours of each machine.

Typical column headings used in the tabulation for allocation of factory costs in machine rate determination are as follows:

1. Present value of machine
2. Remaining life of machine (hours)
3. Hours per year
4. Machines-per-man
5. Labor rate
6. Floor area of machine
7. Depreciation per hour

8. Depreciation per hour of nonproductive equipment
9. Supervision
10. Shipping and receiving
11. Inspection
12. Tools and supplies
13. Building costs
14. Power and light
15. Maintenance of tools
16. Maintenance and repair of equipment
17. Property taxes and insurance
18. Miscellaneous
19. Direct labor
20. Fringe benefits
21. Total of 7 through 20 (this is the machine rate)

Definitions of the above elements of factory cost are as follows:

1. *Present value.* Estimated cost of replacement with a used machine in approximately the same condition as the machine in question. An appraisal by a used equipment dealer is recommended. For a new machine, installed cost should be used initially, less about 15%.
2. *Remaining life.* Estimate of the number of hours the machine can be expected to operate before it will have to be replaced or undergo a major rebuilding job.
3. *Hours per year.* Number of hours each machine can be expected to operate, from historical data, modified to what is considered to be normal operating level for the coming year.
4. *Machines-per-man.* Long-term average machines-per-man ratios should be estimated for equipment normally operated in complements of two or more machines.
5. *Labor rate.* Operator straight time hourly rate for the machine in question. Normally the top-rated operators' rates are used.

6. *Floor area of machine.* Actual area of each machine in square feet. No allowance is made for additional floor area for storage of work, supporting equipment, etc. It is assumed for this purpose that such additional space is proportional to the floor area of the machine.
7. *Depreciation per hour.* Column 1 divided by column 2 for the machine in question.
8. *Depreciation per hour of nonproductive equipment.* From appraisal determine the present value of such equipment (tool sharpening, inspection, material handling, etc.) as a percentage of the value of production equipment (call it *X*%). Assume it has an average remaining life the same as production equipment and assign a depreciation figure of *X*% of column 7 for machine in question.
9. *Supervision.* Estimate how much supervision time is spent on an average with the type of machine in question. The extended sum should equal projected total supervision cost per year.
10. *Shipping and receiving.* Divide projected total annual cost by total hours per year. Allocate equally to each machine.
11. *Inspection.* Estimate in same manner as for column 9. Total should equal projected annual inspection cost.
12. *Tools and supplies.* Estimate in same manner as for columns 9 and 11. Total should equal projected annual expenditure for tools and supplies.
13. *Building costs.* Building depreciation, rent, heat, building maintenance (including sweeping), property taxes, and insurance. Allocate these costs proportionately to area of machine and hours worked per machine.
14. *Power and light.* Allocate total power and light cost proportionately to connected horsepower of each machine.
15. *Maintenance of tools.* This includes tool crib labor, tool sharpening, and repair and maintenance of jigs and fixtures. Express the total as a percentage of column 12 total (call it *Y*%). This allocation is then *Y*% of column 12 for each machine.

16. *Maintenance and repair of equipment.* Allocate from historical data.
17. *Property taxes and insurance.* Allocate total annual cost in proportion to present value of each machine and hours worked by each machine.
18. *Miscellaneous.* All other shop overhead items not included in any of the foregoing categories. Divide this total by total hours per year and allocate equally to each machine.
19. *Direct labor.* Column 5 divided by column 4 times a factor that corrects for overtime premium, nightshift differential, leadman rates that are higher than for the top-rated operators, and rates that are lower than for the top-rated operators.
20. *Fringe benefits.* The total of all fringe benefits (vacations, holidays, hospitalization, pensions, unemployment contribution, payroll taxes, social security, etc.). This total should include fringe benefits for indirect as well as direct labor, and can be expressed as a percentage of direct labor. The fringe benefit allocation is then this percentage of column 19 for each machine.
21. *The sum of columns 7 through 20 for each machine.* This total is the machine-hour rate.

The extended totals for all machines of variable expense items (columns 9, 10, 11, 12, 15, 16, 19, and 20) serves as a basis for budgetary control. If, say, the machine-hours projected for the next year, or next quarter, or next month were 80% of the operating level used in setting the rates, then the budget "aim" for the total of variable expense items should be no more than 80% of the extended total for the operating level used in setting the rates.

Since a machine-hour rate cost system is developed from projected operating levels and machines-per-man ratios, "variances" result that increase or decrease as the degree of departure from projected values changes. If, for example, the total productive machine-hours worked in a given month, converted into dollars by multiplying by the various machine-hour rates, is less than the actual direct labor plus burden for that month, the difference is called a variance, and in this case would be negative. In such a case labor and burden is said to be "underabsorbed." In the opposite situation labor and burden is "overabsorbed."

The variance described is actually a composite of two variances, and for control purposes these should be examined separately each month. One is an "efficiency variance," which is the difference between actual and projected machines-per-man ratios. The second is a "production variance," which results from operating either more or fewer machine-hours than projected, from a variation in mix of work that over- or underutilizes certain equipment, and from spending over- or underbudgeted expense items.

This periodic study of variances, in addition to its importance for control purposes, provides a means for testing the rate system. If variances have a trend in one direction, it could be a sign that the rates should be redeveloped.

There is of course another variance that must be monitored all the time. This is the "performance variance," the difference between standard and actual machine-hours.

To convert present value to book value depreciation, believed desirable to satisfy auditors because machines that are written off are given a present value, which could result in total depreciation being higher than book, or lower if there are few such machines.

This conversion can be accomplished by restating item 7 as follows:

"Depreciation per hour—Column 1 (adjusted to book) divided by column 2 for each machine. Adjust column 1 as follows:

> Step 1—Total the book values of all production equipment.
>
> Step 1—Calculate the percentage of the present value of each machine to the total of all the present values, then apply these percentages to the step 1 total. This provides the depreciation per hour for each production machine."

Item 8 remains as stated. With the above changes in item 7, the totals of items 7 and 8 will equal the book depreciation. Present values are still used but only as an allocation formula.

ACTIVITY-BASED COSTING

We have been discussing factory costs related to the machining of gears. At some gear manufacturers there are significant costs incurred that are not directly related to the machining of gears, and whose inclusion in a machine-hour rate could cause distortions in a finished gear's cost of manufacture.

An example is General, Selling, and Administrative expenses (GS&A). In a small gear shop with relatively small GS&A costs it may be appropriate to include these expenses as part of the machine-hour rate or as a single markup to the factory cost of a product. In some situations, however, such general treatment would distort a product's cost.

For example, if half of a company's sales came from an old, established, customer, and required little or no marketing effort, while the other half is gained through aggressive (and expensive) sales efforts, it would be unfair to charge these costs to all products through either the machine-hour rates or through a single markup. These sales costs should be attributed only to the half of the products to which they apply, not to the old, established customer's products.

Gear manufacturers who incur significant costs in manufacturing operations that are not directly related to the machining of gears, or whose GS&A expenses are either large or do not apply equally to all products, should consider using the concept of Activity-Based Costing ("ABC") to help develop a more equitable way to allocate these costs in proportion to the activity required for each product.

A good reference for "ABC" is a book entitled *Activity-Based Costing for Small and Mid-Sized Businesses: An Implementation Guide,* by Douglas T. Hicks, John Wiley & Sons, Inc., New York, 1992.

17

Drafting Practice—How to Make Gear Drawings

This chapter is intended to instruct those who use gears how to make gear drawings.

For the gear designer who is a gear user, not a gear maker, who is using his or her drawings to request bids, less detail is required than on drawings for manufacture by the gear maker, and the quality required needs to be designated only by use of AGMA Quality Numbers. Sufficient detail is required, however, for use by the gear users' receiving inspectors.

We start by showing how the spur gear and pinion described in Chapter 12 would look when they are drawn using typical drafting practice. The gear is shown made from a weldment as well as from a casting.

A helical gear would look the same except the casting or weldment design would include stiffener ribs on both sides of each spoke, or on both sides of the web of the weldment between the lightening holes in the web. Stiffener ribs on helical gears are desirable to withstand the lateral thrust characteristic of helical gearing.

Other spoke or web designs include "I" section or "H" section configurations for wide face gears, to provide structural rigidity to the gear blanks.

In addition to the data shown for the gears on pp. 146 and 147, each of these should also show number of teeth, pitch, pressure angle, and AGMA Quality Number. Also the drawing for the bevel gear made from a drop forging (Figure 17.6) shows that the tapered hub OD and

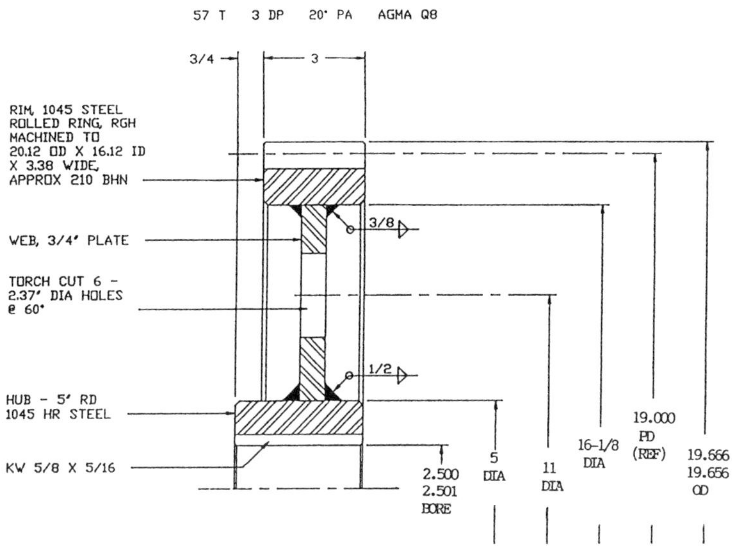

Figure 17.1. Spur gear. This drawing shows the gear described in Chapter 12, p. 102 made from a weldment.

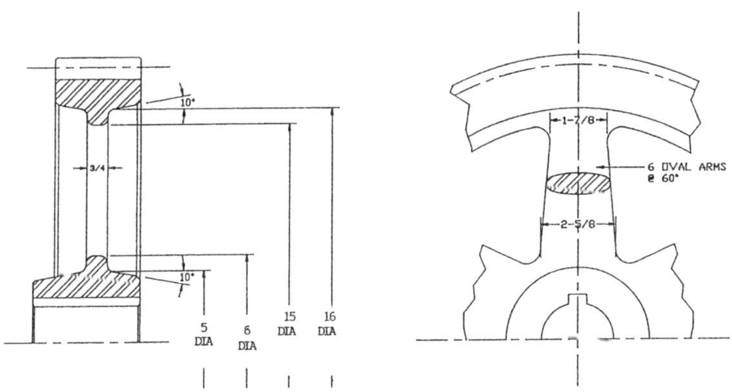

Figure 17.2. This drawing shows the gear described in Chapter 12, p. 102, made from a casting. Dimensions omitted are the same as the weldment drawing above.

the back surface of the web are not to be finish machined but should have the as-forged finish.

In Figure 17.7 the type of bronze must be specified for the worm gear ring, and for the bevel gears referred to above, material and heat treat specifications must be given.

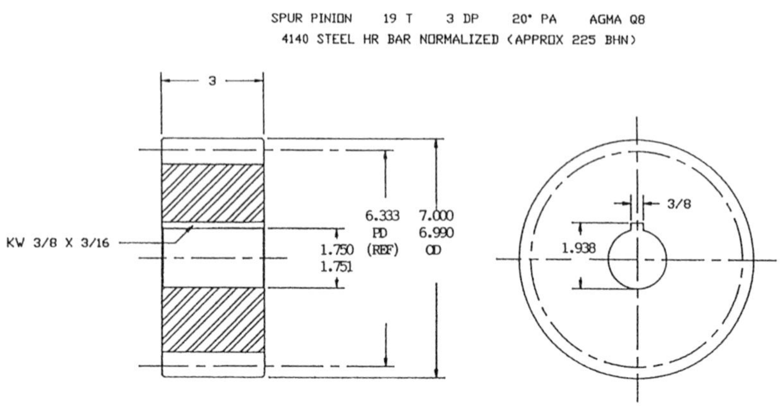

Figure 17.3. This drawing shows the pinion described in Chapter 12, p. 102. Bore diameter is assumed, 4140 Steel is selected as it has greater work hardening capabilities than carbon steel. The face width is dimensioned 3″, the same as the mating gear. Often with open gearing the pinion would be specified to have a face ⅛″ to ¼″ wider than its mate so lateral positioning in assembly is less critical.

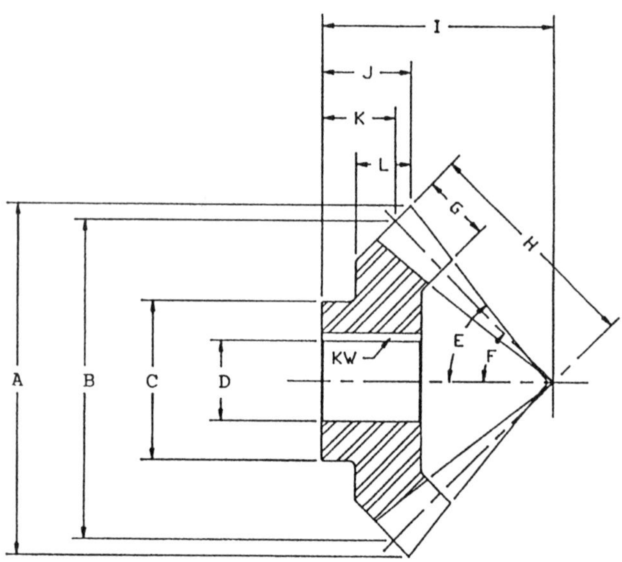

Figure 17.4. Miter gear (1.1 ratio bevel gearing). A, Outside diameter; B, pitch diameter; C, hub diameter; D, bore diameter; E, face angle; F, pitch angle (45° for miters); G, face width; H, cone distance; I, mounting distance; J, backing to point; K, backing to pitch line; L, point to back face.

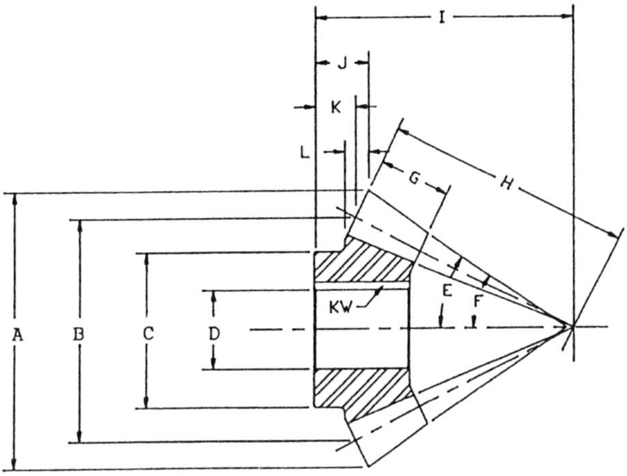

Figure 17.5. Bevel pinion. A, Outside diameter; B, pitch diameter; C, hub diameter; D, bore diameter; E, face angle; F, pitch angle; G, face width; H, cone distance; I, mounting distance; J, backing to point; K, backing to pitch line; L, point to back face.

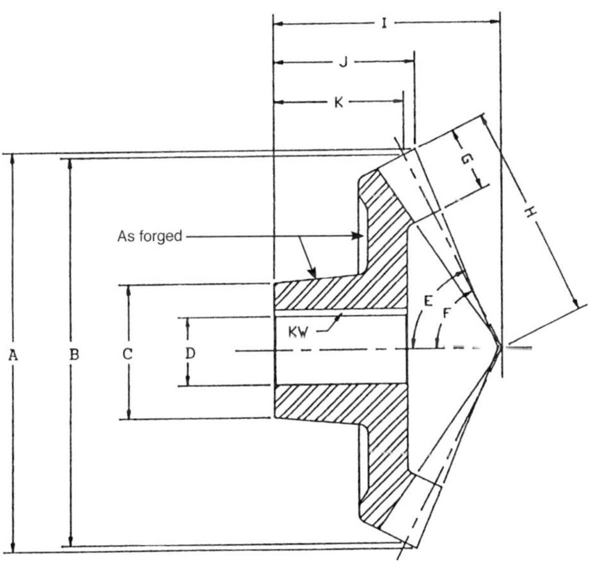

Figure 17.6. Bevel gear—made from drop forging. A, Outside diameter; B, pitch diameter; C, hub diameter; D, bore diameter; E, face angle; F, pitch angle; G, face width; H, cone distance; I, mounting distance; J, backing to point; K, backing to pitch line.

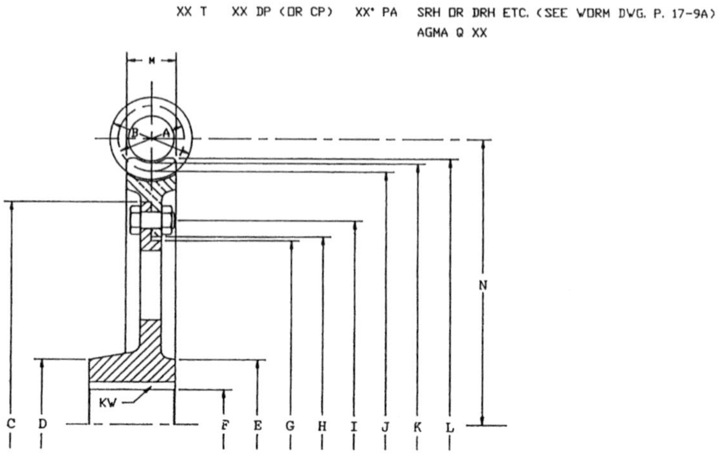

Figure 17.7. Bronze worm gear mounted on cast iron center. A, PD of worm; B, OD of worm; C, OD of cast iron worm gear center and diameter of bronze worm gear counterbore, slip fit; D, hub diameter projecting end; E, hub diameter flush end; F, bore diameter; G, clearance diameter cast iron center; H, clearance diameter bronze worm gear id; I, diameter of bolt circle; J, PD of worm gear (ref); K, throat diameter of worm gear; L, OD of worm gear; M, face width of worm gear; N, center distance.

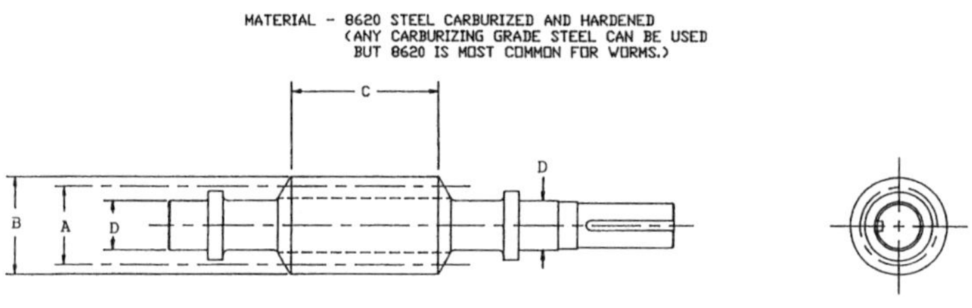

Figure 17.8. Worm integral with shaft. A, PD of worm; B, OD of worm; C, face width of worm; D, Brg dias—tolerances per bearing catalog. As shown above following description it is necessary to specify number of threads and hand, pitch, pressure angle and AGMA quality number. The number of threads and hand are abbreviated as follows.

Single thread right hand: srh (or lh); Double thread right hand: drh (or lh); Triple thread right hand: trh (or lh). Four, five, and more threads are usually indicated by the numerals; thus 4 thread would be 4 thrd. rh (or lh).

The foregoing drawings cover most, if not all, of the most common gear types. It is believed that sufficient detail is provided to instruct the reader on the fundamentals of making gear drawings.

18

Gear Technology Developments, 1960 to the Present and Beyond

As Mr. Dudley pointed out in *The History of Gears* at the start of this book, the advances in gear technology and manufacture matured in the period 1930–1960, and even more greatly in the 1980s. So first we will talk about the important developments during the 1960s and the 1980s.

The 1960s

In the 1960s AGMA further developed standards for the rating of surface durability and strength of spur, helical, and bevel gear teeth.

Metallurgical quality gained importance in gear design considerations during this period.

Gear honing machines were developed to provide better finish than grinding, required on high-speed highly loaded gears in the aerospace field.

New designs of cutting tools and special materials for cutting tool teeth made possible the cutting of harder materials. One of these was the "skiving" hob which can cut gears up to about 60 RC hardness.

EP lubricants were improved and synthetic oils came into use for high-temperature applications.

The 1980s

"CBN" grinding was developed for high-volume work. CBN stands for cubic boron nitride, an abrasive used as a coating on metal wheels.

Higher speed grinding with vitreous wheels was developed.

A new cast iron called "austembered ductile iron" (ADI) was being developed. This material has good fatigue and toughness characteristics, is 10% lighter in weight than steel, has improved noise and vibration dampening, and has improved wear and scuffing resistance.

Gear lubricants were further improved.

There were more developments of honing techniques to provide better finishes than grinding.

CNC hobbing and shaping machines were developed that, among other advantages, reduce setup times substantially. CNC stands for "Computer Numerical Control." Earlier numerical control machines (NC) were controlled by tape.

Titanium coating for hobs, shaper cutters, and other cutting tools was developed. Coated tools will have a longer life and can operate at faster cutting speeds than tools that are not coated.

The Early 1990s

There were developments in cutting tools. Two examples were developed by Pfauter-Maag Cutting Tools of Loves Park, IL. One is called the Wafer™ shaper cutter, the other the Wafer™ hob (the hob has no wafers but is so named as it was developed for the same purpose as the shaper cutter—to be thrown away when worn out, not resharpened). Illustrations of these tools appear in Figures 18.1 and 18.2. These were reproduced from papers by Geoffrey Ashcroft and Brian W. Cluff (see References)

The Wafer shaper cutter uses a thin disc with teeth to provide the cutting edges. It will last longer than a single sharpening of a conventional cutter. When it becomes dull it is thrown away and replaced with a new wafer. No changes in machine settings are required, as is the case with a conventional cutter every time it is resharpened.

The Wafer hob is longer than a conventional hob. Instead of resharpening it is shifted to a new cutting position, and after several shifts it is thrown away. It has many more gashes than a conventional hob, hence has much thinner teeth which cannot be resharpened. The additional gashes produce a smoother tooth surface than can be obtained with a conventional hob.

There are limitations in application of these hobs. They are, however,

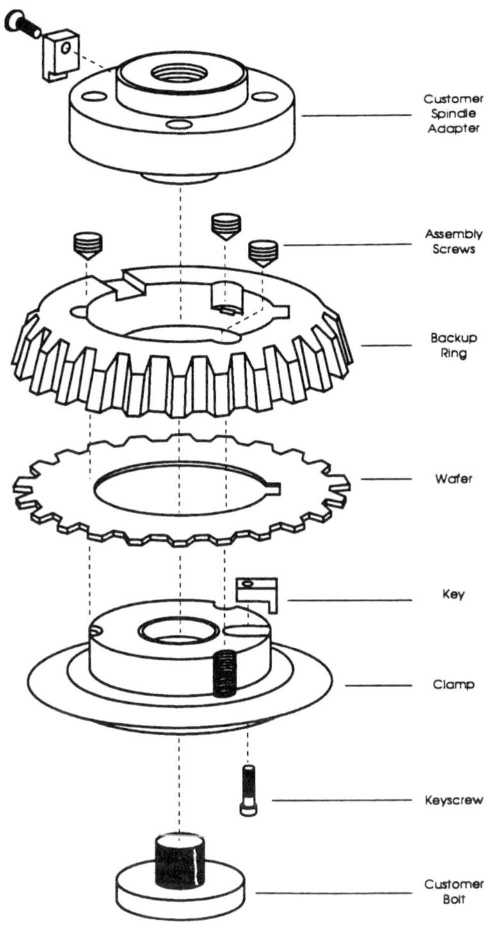

Figure 18.1 Exploded view of a wafer™ shaper cutter assembly.

an attractive solution to new production applications in relatively fine pitches of 10 DP and finer.

1992 and Beyond

It is always difficult to predict what might happen in the future. It is safe to say, however, that we can expect more developments. Some of these will be:

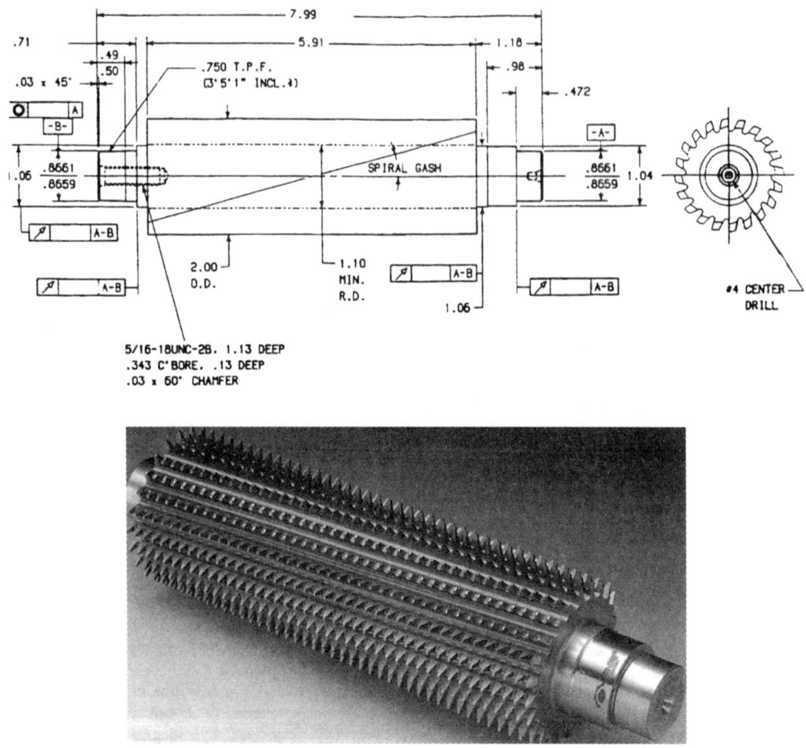

Figure 18.2. Design of a helically gashed shank type solid Wafer™ hob for 10 DP or finer.

Further improvements in gear rating formulas.

New gear materials, both metallic and nonmetallic will come into use.

Gear lubricant technology will make further progress, both for oil used and for the additives.

There will be further improvement in gear quality and methods of quality control.

For high production gears the trend will continue toward full automation of all machining and inspection operations.

Honing, peening, plating, laser bombardment, etc. can be expected to expand further in the next 20 or so years.

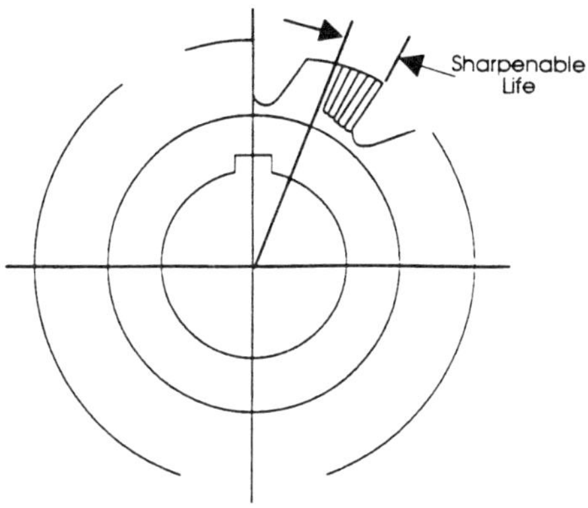

Figure 18.3. Sharpenable life of a hob.

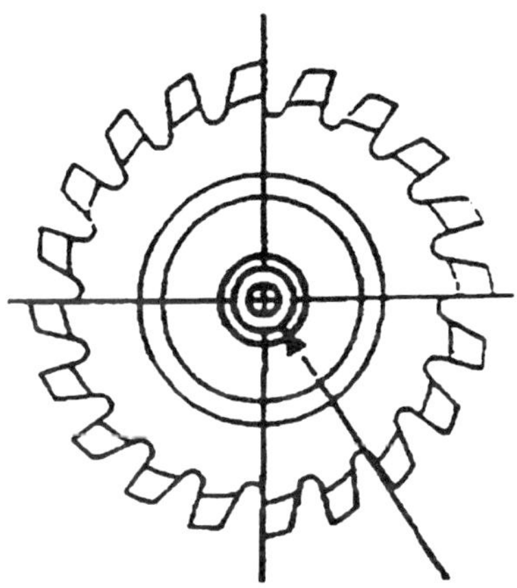

Figure 18.4. Nonresharpenable Wafer^{TM} hob

To summarize, in the words of Mr. Dudley in the last paragraph of his paper:*

**Gear Technology . . . Past, Present and Future,* by Darle W. Dudley, a paper presented to the International Conference on Gearing, in Zhengzhow, Henan, China, 1988.

"Gears are here to stay. All manner of machinery on land, in water, in the air and in outer space require gears. The weight, cost reliability and excellence of performance (in terms of noise and reliability) are of prime importance. The development of gear design and the development of ways to make gears will go on indefinitely. Man's progress on earth and in the whole cosmos beyond the earth requires a continued evolution of the gear art."

19

Recent Developments in Gear Cutting Machines

A few of the recent developments in gear cutting machines are described and illustrated in this chapter.

Fellows Corporation has developed and is producing a line of CNC Hydrostroke® Gear Shapers. The heart of these machines is a hydromechanical system that applies the entire cutting force concentric to the cutting spindle, eliminating the side loads that occur in crank-type stroking systems. They also feature "rapid return," the noncutting portion of a stroke cycle, which increases production. The CNC system provides faster setup times and permits virtually unattended operation. Such a machine is illustrated in Figure 19.1. Fellows has developed many other features on the equipment they produce.

The Gleason Works has developed and is now producing a new line of spiral bevel and hypoid gear generating and tooth grinding machines, called the Phoenix®. These machines use six-axis CNC to produce spiral bevel and hypoid gear geometrics with no mechanical adjustments and make faster setups.

Gleason also now manufactures a line of hobbing machines, called the G-Tech™ Machines. These are CNC machines, eliminating the need for index and differential change gears, manual setup of hob swivel angles, hob position, and hob feed. Also crown and taper hobbing can be performed without the use of cams. These machines are illustrated in Figures 19.2 and 19.3.

National Broach & Machine Co. has further developed a number of their Red Ring® line of machines. One is the model GFD single axis CNC rotary gear shaving machine, which can conventional, diagonal,

Figure 19.1. Fellows CNC Hydrostroke® Gear Shaper. (Courtesy of Fellows Corporation, Springfield, VT.)

Figure 19.2. Gleason Phoenix® Spiral Bevel and Hypoid Generator. (Courtesy of The Gleason Works, Rochester, NY.)

Figure 19.3. Gleason G-Tech® Hobbing Machine. (Courtesy of The Gleason Works, Rochester, NY.)

plunge, or tangential shave, with crowning taper or straight attachments. Another is the Red Ring CNC 5 axis GF-300 gear finisher. This machine can be configured as a shaver, hard honer, roll finisher, and CBN (cubic boron nitride) form grinder. Still another is the Red Ring SF-900 CNC precision gear grinder. This machine features versatility in choosing the best grinding wheel for the job, fast setups, automatic wheel size compensation, and stock dividing. The GF-300 and SF-900 machines are illustrated in Figures 19.4 and 19.5. Other National Broach products include broaching machines and many others, all of which have been improved and/or redesigned over recent years.

American Pfauter has developed and is producing their latest vertical hobbing machines, the PE150 and PE300. Both are CNC machines. They feature shorter machining times, faster hob change, lower tool costs, shorter idle times (when one man is operating more than one machine), greater versatility, and faster setup. Both machines are available with automatic loading and unloading systems for production jobs. The PE150 is illustrated in Figure 19.6. This machine is equipped with a loading and unloading system.

Figure 19.4. National Broach Red Ring® CNC GF-300 Gear Finisher. (Courtesy of National Broach & Machine Co., Mt. Clemens, MI.)

Figure 19.5. National Broach Red Ring® SF-900 Precision Gear Grinder. (Courtesy of National Broach & Machine Co., Mt. Clemens, MI.)

Figure 19.6. American Pfauter PE 150 CNC Hobbing Machine. (Courtesy of American Pfauter, Loves Park, IL.)

There are of course other gear machine manufacturers who have done a lot of development work, but space does not permit mentioning all of them here. AGMA can give you their names and addresses.

Appendix A
AGMA Standards—A Partial List

This is a partial list of AGMA Standards, selected for use by those who wish to pursue gear design procedures.

Number	*Title*	*List Price**
390.03	Handbook—Gear Classification, Materials, and Measuring Methods for Bevel, Hypoid, Fine Pitch Worm Gearing and Racks, Only as Unassembled Gears	$ 35.00
908-B89	Geometry Factors for Determining the Pitting Resistance and Bending Strength of Spur, Helical and Herringbone Teeth	$ 60.00
1012-F90	Gear Nomenclature, Definitions of Terms, with Symbols	$ 45.00
2000-A88	Gear Classification and Inspection Handbook—Tolerances and Measuring Methods for Unassembled Spur and Helical Gears	$100.00
2001-C95	Fundamental Rating Factors and Calculation Methods for Involute Spur and Helical Gear Teeth	$150.00
2002-B88	Tooth Thickness Specification and Measurement	$ 75.00
2003-A86	Rating the Pitting Resistance and Bending Strength of Generated Straight Bevel, ZEROL Bevel and Spiral Bevel Gear Teeth	$100.00
2004-B89	Gear Materials and Heat Treat Manual	$ 85.00
2005-B88	Design Manual for Bevel Gears	$ 90.00
2008-B90	Standard for Assembling Bevel Gears	$ 55.00
6022-C93	Design Manual for Cylindrical Wormgearing	$ 65.00
6030-C87	Design of Industrial Double-Enveloping Worm-gears	$ 50.00
9005-D94	Industrial Gear Lubrication	$ 70.00

*List price is for non-members of AGMA, effective 1995.

A complete list of standards can be obtained from AGMA at no charge. Address: 1500 King St., Suite 201, Alexandria, VA 22314.

Appendix B
Medium and Coarse Pitch Tolerances for Spur, Helical, and Herringbone Gears

Table A-1.

(Tolerances in Ten-Thousandths of an Inch)

AGMA Quality Number	Normal diametral pitch	Runout tolerance							
		Pitch diameter (inches)							
		3/4	1½	3	6	12	25	50	100
3	1/2					788.2	938.6	1106.9	1305.5
	1				477.8	563.5	671.1	791.4	933.4
	2			289.7	341.6	402.9	479.8	565.9	667.4
	4			207.1	244.3	288.1	343.1	404.6	477.2
	8			148.1	174.7	206.0	245.3	289.3	341.2
4	1/2					563.0	670.4	790.7	932.5
	1				341.3	402.5	479.3	565.3	666.7
	2			206.9	244.0	287.8	342.7	404.2	476.7
	4			147.9	174.5	205.8	245.0	289.0	340.8
	8			105.8	124.8	147.1	175.2	206.6	243.7
5	1/2					402.1	478.9	564.8	666.1
	1				243.8	287.5	342.4	403.8	476.2
	2			147.8	174.3	205.6	244.8	288.7	340.5
	4		89.6	105.7	124.6	147.0	175.0	206.4	243.5
	8		64.1	75.6	89.1	105.1	125.1	147.6	174.1
6	1/2					287.2	342.1	403.4	475.8
	1				174.1	205.4	244.6	288.4	340.2
	2			105.6	124.5	146.8	174.9	206.2	243.2
	4		64.0	75.5	89.0	105.0	125.0	147.4	173.9
	8	38.8	45.8	54.0	63.6	75.1	89.4	105.4	124.3
	12	31.9	37.6	44.4	52.3	61.7	73.5	86.6	102.2
	20	24.9	29.4	34.6	40.8	48.2	57.4	67.7	79.8
7	1/2					205.2	244.3	288.1	339.8
	1				124.4	146.7	174.7	206.0	243.0
	2			75.4	88.9	104.9	124.9	147.3	173.7
	4		45.7	53.9	63.6	75.0	89.3	105.3	124.2
	8	27.7	32.7	38.5	45.5	53.6	63.9	75.3	88.8
	12	22.8	26.9	31.7	37.4	44.1	52.5	61.9	73.0
	20	17.8	21.0	24.7	29.2	34.4	41.0	48.3	57.0

Runout tolerance Pitch diameter (inches) 200	Runout tolerance Pitch diameter (inches) 400	Pitch tolerance Pitch diameter (inches) 3/4	1½	3	6	12	25	50	100	200	400
1539.6	1815.7										
1100.8	1298.1										
787.1	928.2										
562.7	663.7										
402.4	474.5										
1099.7	1297.0										
786.3	927.3										
562.2	663.0										
402.0	474.1										
287.4	338.9										
785.5	926.4										
561.6	662.4										
401.6	473.6										
287.1	332.6										
205.3	242.1										
561.1	665.7					38.4	43.7	49.4	55.9	63.2	71.4
401.2	473.1				29.1	32.9	37.4	42.3	47.8	54.1	61.1
286.8	338.3			22.0	24.9	28.1	32.0	36.2	41.0	46.3	52.3
205.1	241.9		16.7	18.9	21.3	24.1	27.4	31.0	35.1	39.6	44.8
146.6	272.9	12.6	14.3	16.1	18.2	20.6	23.5	26.6	30.0	33.9	38.4
120.5	242.1	11.5	13.0	14.7	16.7	18.8	21.5	24.3	27.4	31.0	35.0
94.1	111.0	10.3	11.6	13.1	14.9	16.8	19.1	21.6	24.5	27.6	31.3
400.8	472.7					27.0	30.8	34.8	39.3	44.5	50.3
286.6	337.9				20.5	23.1	26.4	29.8	33.7	38.1	43.1
204.9	241.6			15.5	17.5	19.8	22.6	25.5	28.8	32.6	36.9
146.5	172.8		11.7	13.3	15.0	17.0	19.3	21.8	24.7	29.9	31.6
104.7	123.5	8.9	10.1	11.4	12.9	14.5	16.5	18.7	21.1	23.9	27.0
86.1	101.5	8.1	9.2	10.4	11.7	13.3	15.1	17.1	19.3	21.8	24.7
67.2	79.3	7.2	8.2	9.3	10.5	11.8	13.5	15.2	17.2	19.5	22.0

Table A-1. (Continued)
(Tolerances in Ten-Thousandths of an Inch)

AGMA Quality Number	Normal diametral pitch	Runout tolerance										Pitch tolerance				
		Pitch diameter (inches)										Pitch diameter (inches)				
		3/4	1½	3	6	12	25	50	100	200	400	3/4	1½	3	6	12
8	1/2					146.5	174.5	205.8	242.7	286.3	337.6					19.0
	1				88.8	104.8	124.8	147.2	173.6	204.7	241.4				14.4	16.3
	2			53.9	63.5	74.9	89.2	105.2	124.1	146.3	172.6			10.9	12.3	14.0
	4		32.7	38.5	45.4	53.6	63.8	75.2	88.7	104.6	123.4		8.3	9.3	10.6	11.9
	8	19.8	23.3	27.5	32.5	38.3	45.6	53.8	63.4	74.8	88.2	6.3	7.1	8.0	9.0	10.2
	12	16.3	19.2	22.6	26.7	31.5	37.5	44.2	52.1	61.5	72.5	5.7	6.5	7.3	8.3	9.3
	20	12.7	15.0	17.7	20.8	24.6	29.3	34.5	40.7	48.0	56.6	5.1	5.8	6.5	7.4	8.3
9	1/2					104.7	124.7	147.0	173.4	204.5	241.2					13.4
	1				63.5	74.8	89.1	105.1	124.0	146.2	172.4				10.2	11.5
	2			38.5	45.4	53.5	63.7	75.2	88.6	104.5	123.3			7.7	8.7	9.8
	4		23.3	27.5	32.4	38.3	45.6	53.7	63.4	74.7	88.1		5.8	6.6	7.4	8.4
	8	14.1	16.7	19.7	23.2	27.4	32.6	38.4	45.3	53.4	63.0	4.4	5.0	5.6	6.4	7.2
	12	11.6	13.7	16.2	19.1	22.5	26.8	31.6	37.2	43.9	51.8	4.0	4.6	5.1	5.8	6.6
	20	9.1	10.7	12.6	14.9	17.6	20.9	24.7	29.1	34.3	40.4	3.6	4.1	4.6	5.2	5.9
10	1/2					74.8	89.0	105.0	123.8	146.1	172.3					9.4
	1				45.3	53.5	63.7	75.1	88.5	104.4	123.2				7.2	8.1
	2			27.5	32.4	38.2	45.5	53.7	63.3	74.7	88.1			5.4	6.1	6.9
	4		16.7	19.6	23.2	27.3	32.5	38.4	45.3	53.4	63.0		4.1	4.6	5.2	5.9
	8	10.1	11.9	14.0	16.6	19.5	23.3	27.4	32.4	38.2	45.0	3.1	3.5	4.0	4.5	5.1
	12	8.3	9.8	11.5	13.6	16.1	19.1	22.6	26.6	31.4	37.0	2.8	3.2	3.6	4.1	4.6
	20	6.5	7.6	9.0	10.6	12.5	14.9	17.6	20.8	24.5	28.9	2.5	2.9	3.2	3.7	4.1
11	1/2					53.4	63.6	75.0	88.5	104.3	123.0					6.6
	1				32.4	38.2	45.5	53.6	63.2	74.6	88.0				5.0	5.7
	2			19.6	23.1	27.3	32.5	38.3	45.2	53.3	62.9			3.8	4.3	4.9
	4		11.9	14.0	16.6	19.5	23.2	27.4	32.3	38.1	45.0		2.9	3.3	3.7	4.2
	8	7.2	8.5	10.0	11.8	14.0	16.6	19.6	23.1	27.3	32.2	2.2	2.5	2.8	3.2	3.6
	12	5.9	7.0	8.2	9.7	11.5	13.7	16.1	19.0	22.4	26.4	2.0	2.3	2.6	2.9	3.3
	20	4.6	5.5	6.4	7.6	9.0	10.7	12.6	14.8	17.5	20.6	1.8	2.0	2.3	2.6	2.9
12	1/2					38.1	45.4	5.36	63.2	74.5	87.9					4.7
	1				23.1	27.3	32.5	38.3	45.2	53.3	62.8				3.5	4.0
	2			14.0	16.5	19.5	23.2	27.4	32.3	38.1	44.9			2.7	3.0	3.4
	4		8.5	10.0	11.8	13.9	16.6	19.6	23.1	27.2	32.1		2.0	2.3	2.6	2.9
	8	5.2	6.1	7.2	8.5	10.0	11.9	14.0	16.5	19.5	23.0	1.5	1.7	2.0	2.2	2.5
	12	4.2	5.0	5.9	6.9	8.2	9.8	11.5	13.6	16.0	18.9	1.4	1.6	1.8	2.0	2.3
	20	3.3	3.9	4.6	5.4	6.4	7.6	9.0	10.6	12.5	14.7	1.3	1.4	1.6	1.8	2.0

Pitch tolerance					Profile tolerance										Lead tolerance				
Pitch diameter (inches)					Pitch diameter (inches)										Face width (inches)				
25	50	100	200	400	3/4	1½	3	6	12	25	50	100	200	400	1 and Less	2	3	4	5
21.7	24.5	27.7	31.3	35.4					42.6	47.7	53.1	59.1	65.7	73.1	5	8	11	13	16
18.6	21.0	23.7	26.8	30.3				28.3	31.5	35.3	39.3	43.7	48.6	54.1					
15.9	18.0	20.3	23.0	26.0			18.8	21.0	23.3	26.1	29.0	32.3	36.0	40.0					
13.6	15.4	17.4	19.7	22.2		12.5	13.9	15.5	17.2	19.3	21.5	23.9	26.6	29.6					
11.7	13.2	14.9	16.8	19.0	8.3	9.3	10.3	11.5	12.8	14.3	15.9	17.7	19.7	21.9					
10.6	12.0	13.6	15.4	17.4	7.0	7.8	8.6	9.6	10.7	12.0	13.3	14.8	16.5	18.4					
9.5	10.7	12.1	13.7	15.5	5.6	6.2	6.9	7.7	8.6	9.6	10.7	11.9	13.2	14.7					
15.3	17.3	19.5	22.1	24.9					30.4	34.1	37.9	42.2	46.9	52.2	4	7	9	11	13
13.1	14.8	16.7	18.9	21.4				20.2	22.5	25.2	28.1	31.2	34.7	38.6					
11.2	12.7	14.3	16.2	18.3			13.5	15.0	16.7	18.6	20.7	23.1	25.7	28.6					
9.6	10.8	12.2	13.8	15.7		8.9	10.0	11.1	12.3	13.8	15.3	17.1	19.0	21.1					
8.2	9.3	10.5	11.9	13.4	5.9	6.6	7.4	8.2	9.1	10.2	11.4	12.6	14.1	15.6					
7.5	8.5	9.6	10.8	12.2	5.0	5.5	6.2	6.9	7.6	8.6	9.5	10.6	11.8	13.1					
6.7	7.6	8.6	9.7	10.9	4.0	4.4	4.9	5.5	6.1	6.8	7.6	8.5	9.4	10.5					
10.8	12.2	13.7	15.5	17.6					21.7	24.3	27.1	30.1	33.5	37.3	3	5	7	9	10
9.2	10.4	11.8	13.3	15.0				14.5	16.1	18.0	20.0	22.3	24.8	27.6					
7.9	8.9	10.1	11.4	12.9			9.6	10.7	11.9	13.3	14.8	16.5	18.3	20.4					
6.7	7.6	8.6	9.8	11.0		6.4	7.1	7.9	8.8	9.9	11.0	12.2	13.6	15.1					
5.8	6.5	7.4	8.3	9.4	4.2	4.7	5.3	5.9	6.5	7.3	8.1	9.0	10.0	11.2					
5.3	6.0	6.7	7.6	8.6	3.6	4.0	4.4	4.9	5.5	6.1	6.8	7.6	8.4	9.4					
4.7	5.3	6.0	6.8	7.7	2.9	3.2	3.5	3.9	4.4	4.9	5.4	6.1	6.7	7.5					
7.6	8.6	9.7	10.9	12.4					15.5	17.4	19.3	21.5	24.0	26.7	3	4	6	7	8
6.5	7.3	8.3	9.4	10.6				10.3	11.5	12.9	14.3	15.9	17.7	19.7					
5.6	6.3	7.1	8.0	9.1			6.9	7.6	8.5	9.5	10.6	11.8	13.1	14.6					
4.8	5.4	6.1	6.9	7.8		4.6	5.1	5.6	6.3	7.0	7.8	8.7	9.7	10.8					
4.1	4.6	5.2	5.9	6.6	3.0	3.4	3.8	4.2	4.6	5.2	5.8	6.4	7.2	8.0					
3.7	4.2	4.7	5.4	6.1	2.5	2.8	3.1	3.5	3.9	4.4	4.9	5.4	6.0	6.7					
3.3	3.7	4.2	4.8	5.4	2.0	2.3	2.5	2.8	3.1	3.5	3.9	4.3	4.8	5.4					
5.3	6.0	6.8	7.7	8.7					11.1	12.4	13.8	15.4	17.1	19.0	2	3	5	6	7
4.6	5.2	5.8	6.6	7.5				7.4	8.2	9.2	10.2	11.4	12.7	14.1					
3.9	4.4	5.0	5.6	6.4			4.9	5.5	6.1	6.8	7.6	8.4	9.4	10.4					
3.3	3.8	4.3	4.8	5.5		3.3	3.6	4.0	4.5	5.0	5.6	6.2	6.9	7.7					
2.9	3.2	3.7	4.1	4.7	2.2	2.4	2.7	3.0	3.3	3.7	4.1	4.6	5.1	5.7					
2.6	3.0	3.3	3.8	4.3	1.8	2.0	2.2	2.5	2.8	3.1	3.5	3.9	4.3	4.8					
2.3	2.6	3.0	3.4	3.8	1.5	1.6	1.8	2.0	2.2	2.5	2.8	3.1	3.4	3.8					

Table A-1. (Continued)
(Tolerances in Ten-Thousandths of an Inch)

AGMA Quality Number	Normal diametral pitch	Runout tolerance										Pitch tolerance				
		Pitch diameter (inches)										Pitch diameter (inches)				
		3/4	1½	3	6	12	25	50	100	200	400	3/4	1½	3	6	12
13	1/2					27.2	32.4	38.3	45.1	53.2	62.8					3.3
	1				16.5	19.5	23.2	27.4	32.3	38.1	44.9				2.5	2.8
	2			10.0	11.8	13.9	16.6	19.6	23.1	27.2	32.1			1.9	2.1	2.4
	4		6.1	7.2	8.4	10.0	11.9	14.0	16.5	19.5	22.9		1.4	1.6	1.8	2.1
	8	3.7	4.3	5.1	6.0	7.1	8.5	10.0	11.8	13.9	16.4	1.1	1.3	1.4	1.6	1.8
	12	3.0	3.6	4.2	5.0	5.9	7.0	8.2	9.7	11.4	13.5	1.0	1.1	1.3	1.4	1.6
	20	2.4	2.8	3.3	3.9	4.6	5.4	6.4	7.6	8.9	10.5	0.9	1.0	1.1	1.3	1.4
14	1/2					19.5	23.2	27.3	32.2	38.0	44.8					2.3
	1				11.8	13.9	16.6	19.5	23.0	27.2	32.1				1.8	2.0
	2			7.2	8.4	9.9	11.8	14.0	16.5	19.4	22.9			1.3	1.5	1.7
	4		4.3	5.1	6.0	7.1	8.5	10.0	11.8	13.9	16.4		1.0	1.1	1.3	1.5
	8	2.6	3.1	3.7	4.3	5.1	6.1	7.1	8.4	9.9	11.7	0.8	0.9	1.0	1.1	1.2
	12	2.2	2.5	3.0	3.5	4.2	5.0	5.9	9.9	8.2	9.6	0.7	0.8	0.9	1.0	1.1
	20	1.7	2.0	2.3	2.8	3.3	3.9	4.6	5.4	6.4	7.5	0.6	0.7	0.8	0.9	1.0
15	1/2					13.9	16.6	19.5	23.0	27.2	32.0					1.6
	1				8.4	9.9	11.8	14.0	16.5	19.4	22.9				1.2	1.4
	2			5.1	6.0	7.1	8.5	10.0	11.8	13.9	16.4			0.9	1.1	1.2
	4		3.1	3.7	4.3	5.1	6.1	7.1	8.4	9.9	11.7		0.7	0.8	0.9	1.0
	8	1.9	2.2	2.6	3.1	3.6	4.3	5.1	6.0	7.1	8.4	0.5	0.6	0.7	0.8	0.9
	12	1.5	1.8	2.1	2.5	3.0	3.6	4.2	4.9	5.8	6.9	0.5	0.6	0.6	0.7	0.8
	20	1.2	1.4	1.7	2.0	2.3	2.8	3.3	3.9	4.6	5.4	0.4	0.5	0.6	0.6	0.7

Pitch tolerance					Profile tolerance										Lead tolerance				
Pitch diameter (inches)					Pitch diameter (inches)										Face width (inches)				
25	50	100	200	400	3/4	1½	3	6	12	25	50	100	200	400	1 and Less	2	3	4	5
3.8	4.2	4.8	5.4	6.1					7.9	8.9	9.9	11.0	12.2	13.6					
3.2	3.6	4.1	4.6	5.3				5.3	5.9	6.6	7.3	8.1	9.0	10.1					
2.8	3.1	3.5	4.0	4.5			3.5	3.9	4.3	4.9	5.4	6.0	6.7	7.4					
2.4	2.7	3.0	3.4	3.8		2.3	2.6	2.9	3.2	3.6	4.0	4.4	4.9	5.5	2	3	4	4	5
2.0	2.3	2.6	2.9	3.3	1.5	1.7	1.9	2.1	2.4	2.7	3.0	3.3	3.7	4.1					
1.8	2.1	2.4	2.7	3.0	1.3	1.4	1.6	1.8	2.0	2.2	2.5	2.8	3.1	3.4					
1.6	1.9	2.1	2.4	2.7	1.0	1.2	1.3	1.4	1.6	1.8	2.0	2.2	2.5	2.7					
2.6	3.0	3.4	3.8	4.3					5.7	6.3	7.1	7.8	8.7	9.7					
2.3	2.6	2.9	3.3	3.7				3.8	4.2	4.7	5.2	5.8	6.5	7.2					
1.9	2.2	2.5	2.8	3.2			2.5	2.8	3.1	3.5	3.9	4.3	4.8	5.3					
1.7	1.9	2.1	2.4	2.7		1.7	1.9	2.1	2.3	2.6	2.9	3.2	3.5	3.9	1	2	3	4	4
1.4	1.6	1.8	2.1	2.3	1.1	1.2	1.4	1.5	1.7	1.9	2.1	2.3	2.6	2.9					
1.3	1.5	1.7	1.9	2.1	0.9	1.0	1.1	1.3	1.4	1.6	1.8	2.0	2.2	2.4					
1.2	1.3	1.5	1.7	1.9	0.7	0.8	0.9	1.0	1.1	1.3	1.4	1.6	1.8	2.0					
1.9	2.1	2.4	2.7	3.0					4.0	4.5	5.0	5.6	6.2	6.9					
1.6	1.8	2.0	2.3	2.6				2.7	3.0	3.3	3.7	4.1	4.6	5.1					
1.4	1.5	1.7	2.0	2.2			1.8	2.0	2.2	2.5	2.8	3.1	3.4	3.8	1	2	2	3	3
1.2	1.3	1.5	1.7	1.9		1.2	1.3	1.5	1.6	1.8	2.0	2.3	2.5	2.8					
1.0	1.1	1.3	1.4	1.6	0.8	0.9	1.0	1.1	1.2	1.4	1.5	1.7	1.9	2.1					
0.9	1.0	1.2	1.3	1.5	0.7	0.7	0.8	0.9	1.0	1.1	1.3	1.4	1.6	1.7					
0.8	0.9	1.0	1.2	1.3	0.5	0.6	0.7	0.7	0.8	0.9	1.0	1.1	1.3	1.4					

Table A-2. Tooth Thickness Tolerance Classes for Spur, Helical and Herringbone Gearing[2]
(All Tolerance Values in Inches)

Quality Number	Diametral Pitch	Class A	Class B	Class C	Class D	Class E
3	.5	.074				
	1.2	.031				
	2.0	.019				
	3.2	.012				
	5.0	.0075				
4	.5	.074				
	1.2	.031				
	2.0	.019				
	3.2	.012				
	5.0	.0075				
5	.5	.074				
	1.2	.031				
	2.0	.019	.0093			
	3.2	.012	.006			
	5.0	.0075	.0037			
	8.0	.005	.0025			
6	.5	.074				
	1.2	.031				
	2.0	.019	.0093			
	3.2	.012	.006			
	5.0	.0075	.0037			
	8.0	.005	.0025			
	12.0	.003	.0018			
	20.0	.0024	.0012	.0006		
	32.0	.0016	.0008	.00043		
7 and higher	.5	.074				
	1.2	.031				
	2.0	.019	.0093	.0048		
	3.2	.012	.006	.003		
	5.0	.0075	.0037	.0019		
	8.0	.005	.0025	.00125	.00063	
	12.0	.003	.0018	.0009	.00044	
	20.0	.0024	.0012	.0006	.0003	.00016
	32.0	.0016	.0008	.00043	.0002	.0001
	50	.0012	.0006	.0003	.00014	.00007
	80	.0008	.00045	.00022	.00011	.000055
	120	.00067	.00034	.00017	.00009	.000045
	200	.0005	.00025	.00013	.00006	.00003

[2]Extracted from AGMA GEAR HANDBOOK (AGMA 390-03, 1971) with the permission of the publisher, American Gear Manufacturers Association, 1500 King St., Suite 201, Alexandria, VA 22314.

Appendix C

Table A-3.

INTERNAL GEAR DRIVES

PROPORTIONS OF 1 DP PINIONS - 20° FULL DEPTH INVOLUTE BASIC RACK

No. OF TEETH	OUTSIDE RADIUS	PITCH RADIUS	ROOT RADIUS (HOBBED)	ROOT RADIUS (SHAPED)	BASE RADIUS	$\sqrt{R_o^2 - R_b^2}$
N	R_o	R	R_r	R_r	R_b	
16	9.25	8.00	7.093	6.987	7.51754	5.3897
17	9.75	8.50	7.593	7.487	7.98739	5.5914
18	10.25	9.00	8.093	7.988	8.45723	5.7911
19	10.75	9.50	8.593	8.488	8.92708	5.9891
20	11.25	10.00	9.093	8.989	9.39693	6.1854
21	11.75	10.50	9.593	9.489	9.86677	6.3803
22	12.25	11.00	10.093	9.989	10.33667	6.5739
23	12.75	11.50	10.593	10.489	10.80647	6.7662
24	13.25	12.00	11.093	10.989	11.27631	6.9575
25	13.75	12.50	11.593	11.490	11.74616	7.1477
26	14.25	13.00	12.093	11.990	12.21600	7.3370
27	14.75	13.50	12.593	12.490	12.68585	7.5254
28	15.25	14.00	13.093	12.990	13.15570	7.7129
29	15.75	14.50	13.593	13.490	13.62554	7.8998
30	16.25	15.00	14.093	13.991	14.09539	8.0859
31	16.75	15.50	14.593	14.491	14.56524	8.2714
32	17.25	16.00	15.093	14.991	15.03508	8.4562
33	17.75	16.50	15.593	15.491	15.50493	8.6405
34	18.25	17.00	16.093	15.991	15.97477	8.8243
35	18.75	17.50	16.593	16.491	16.44462	9.0076
36	19.25	18.00	17.093	16.992	16.91447	9.1903
37	19.75	18.50	17.593	17.492	17.38431	9.3727
38	20.25	19.00	18.093	17.992	17.85416	9.5546
39	20.75	19.50	18.593	18.492	18.32401	9.7361
40	21.25	20.00	19.093	18.992	18.79385	9.9173
41	21.75	20.50	19.593	19.492	19.26370	10.0981
42	22.25	21.00	20.093	19.992	19.73355	10.2785
43	22.75	21.50	20.593	20.493	20.20339	10.4587
44	23.25	22.00	21.093	20.993	20.67324	10.6385
45	23.75	22.50	21.593	21.493	21.14308	10.8181
46	24.25	23.00	22.093	21.993	21.61293	10.9974
47	24.75	23.50	22.593	22.493	22.08278	11.1764
48	25.25	24.00	23.093	22.993	22.55262	11.3552
49	25.75	24.50	23.593	23.493	23.02247	11.5337
50	26.25	25.00	24.093	23.993	23.49232	11.7121
51	26.75	25.50	24.593	24.493	23.96216	11.8902
52	27.25	26.00	25.093	24.993	24.43201	12.0681
53	27.75	26.50	25.593	25.494	24.90185	12.2458
54	28.25	27.00	26.093	25.994	25.37170	12.4233
55	28.75	27.50	26.593	26.494	25.84155	12.6006
56	29.25	28.00	27.093	26.994	26.31139	12.7778
57	29.75	28.50	27.593	27.494	26.78124	12.9548
58	30.25	29.00	28.093	27.994	27.25109	13.1316
59	30.75	29.50	28.593	28.494	27.72093	13.3083
60	31.25	30.00	29.093	28.994	28.19078	13.4848
61	31.75	30.50	29.593	29.494	28.66062	13.6613
62	32.25	31.00	30.093	29.994	29.13047	13.8375
63	32.75	31.50	30.593	30.494	29.60032	14.0136
64	33.25	32.50	31.093	30.994	30.07014	14.1897
65	33.75	32.50	31.593	31.494	30.54001	14.3655

Table A-3. (Continued)

INTERNAL GEAR DRIVES

PROPORTIONS OF 1 DP PINIONS - 20° FULL DEPTH INVOLUTE BASIC RACK

No. OF TEETH	OUTSIDE RADIUS	PITCH RADIUS	ROOT RADIUS (HOBBED)	ROOT RADIUS (SHAPED)	BASE RADIUS	$\sqrt{R_o^2 - R_b^2}$
N	R_o	R	R_r	R_r	R_b	
66	34.25	33.00	32.093	31.994	31.00986	14.5413
67	34.75	33.50	32.593	32.494	31.47970	14.7170
68	35.25	34.00	33.093	32.994	31.94955	14.8925
69	35.75	34.50	33.593	33.494	32.41940	15.0680
70	36.25	35.00	34.093	33.994	32.88924	15.2433
71	36.75	35.50	34.593	34.494	33.35909	15.4186
72	37.25	36.00	35.093	34.994	33.82893	15.5937
73	37.75	36.50	35.593	35.495	34.29878	15.7688
74	38.25	37.00	36.093	35.995	34.76863	15.9438
75	38.75	37.50	36.593	36.495	35.23847	16.1187
76	39.25	38.00	37.093	36.995	35.70832	16.2935
77	39.75	38.50	37.593	37.495	36.17817	16.4682
78	40.25	39.00	38.093	37.995	36.64801	16.6428
79	40.75	39.50	38.593	38.495	37.11786	16.8174
80	41.25	40.00	39.093	38.995	37.58770	16.9919
81	41.75	40.50	39.593	39.495	38.05755	17.1664
82	42.25	41.00	40.093	39.995	38.52740	17.3407
83	42.75	41.50	40.593	40.495	38.99724	17.5150
84	43.25	42.00	41.093	40.995	39.46709	17.6892
85	43.75	42.50	41.593	41.496	39.93694	17.8634
86	44.25	43.00	42.093	41.996	40.40678	18.0375
87	44.75	43.50	42.593	42.496	40.87663	18.2116
88	45.25	44.00	43.093	42.996	41.34648	18.3856
89	45.75	44.50	43.593	43.496	41.81632	18.5595
90	46.25	45.00	44.093	43.996	42.28617	18.7334
91	46.75	45.50	44.593	44.496	42.75601	18.9073
92	47.25	46.00	45.093	44.996	43.22586	19.0810
93	47.75	46.50	45.593	45.496	43.69571	19.2548
94	48.25	47.00	46.093	45.996	44.16555	19.4285
95	48.75	47.50	46.593	46.496	44.63540	19.6021
96	49.25	48.00	47.093	46.996	45.10525	19.7757
97	49.75	48.50	47.593	47.496	45.57509	19.9492
98	50.25	49.00	48.093	47.996	46.04494	20.1227
99	50.75	49.50	48.593	48.496	46.51478	20.2962
100	51.25	50.00	47.093	48.996	46.98463	20.4696
101	51.75	50.50	49.593	49.496	47.45448	20.6430
102	52.25	51.00	50.093	49.996	47.92432	20.8163
103	52.75	51.50	50.593	50.496	48.39417	20.9896
104	53.25	52.00	51.093	50.996	48.86402	21.1629
105	53.75	52.50	51.593	51.496	49.33386	21.3361
106	54.25	53.00	52.093	51.996	49.80371	21.5093
107	54.75	53.50	52.593	52.496	50.27356	21.6825
108	55.25	54.00	53.093	52.996	50.74340	21.8556
109	55.75	54.50	53.593	53.496	51.21325	22.0287
110	56.25	55.00	54.093	53.996	51.68309	22.2018
111	56.75	55.50	54.593	54.496	52.15294	22.3748
112	57.25	56.00	55.093	54.996	52.62279	22.5478
113	57.75	56.50	55.593	55.497	53.09263	22.7208
114	58.25	57.00	56.093	55.997	53.56248	22.8937
115	58.75	57.50	56.593	56.497	54.03233	23.0666

Table A-4.

INTERNAL GEAR DRIVES

PROPORTIONS OF 1 DP INTERNAL GEARS - 20°FULL DEPTH INVOLUTE BASIC RA

MESHING WITH 16 TOOTH PINIONS AND LARGER

No. OF TEETH	ROOT RADIUS	PITCH RADIUS	INTERNAL RADIUS	BASE RADIUS	$\sqrt{R_i^2 - R_b^2}$	LARGEST CUTTER	LARGEST PINION
N_2	R_r	R	R_i	R_b		N_c	N_1
28	15.472	14.00	13.40	13.15570	2.5470	16	21
29	15.974	14.50	13.90	13.62554	2.7485	16	22
30	16.475	15.00	14.40	14.09539	2.9461	18	23
31	16.977	15.50	14.90	14.56524	3.1406	18	24
32	17.478	16.00	15.40	15.03508	3.3330	20	25
33	17.979	16.50	15.90	15.50493	3.5224	21	26
34	18.480	17.00	16.40	15.97477	3.7103	21	27
35	18.981	17.50	16.90	16.44462	3.8967	21	28
36	19.481	18.00	17.40	16.91447	4.0817	24	29
37	19.982	18.50	17.90	17.38431	4.2656	24	30
38	20.483	19.00	18.40	17.85416	4.4485	24	31
39	20.983	19.50	18.90	18.32401	4.6304	27	32
40	21.484	20.00	19.40	18.79385	4.8117	27	33
41	21.984	20.50	19.90	19.26370	4.9920	27	34
42	22.485	21.00	20.40	19.73355	5.1717	30	35
43	22.985	21.50	20.90	20.20339	5.3510	30	36
44	23.486	22.00	21.40	20.67324	5.5297	30	37
45	23.986	22.50	21.90	21.14308	5.7079	30	38
46	24.487	23.00	22.40	21.61293	5.8857	30	39
47	24.987	23.50	22.90	22.08278	6.0631	30	40
48	25.487	24.00	23.40	22.55262	6.2401	36	41
49	25.988	24.50	23.90	23.02247	6.4168	36	42
50	26.488	25.00	24.40	23.49232	6.5932	36	43
51	26.998	25.50	24.90	23.96216	6.7694	36	43
52	27.489	26.00	25.40	24.43201	6.9452	36	44
53	27.989	26.50	25.90	24.90185	7.1209	36	45
54	28.489	27.00	26.40	25.37170	7.2964	42	46
55	28.989	27.50	26.90	25.84155	7.4716	42	47
56	29.490	28.00	27.40	26.31139	7.6466	42	48
57	29.990	28.50	27.90	26.78124	7.8214	42	49
58	30.490	29.00	28.40	27.25109	7.9961	42	50
59	30.990	29.50	28.90	27.72093	8.1706	42	51
60	31.490	30.00	29.40	28.19078	8.3450	48	52
61	31.990	30.50	29.90	28.66062	8.5193	48	53
62	32.491	31.00	30.40	29.13047	8.6934	48	54
63	32.991	31.50	30.90	29.60032	8.8674	48	55
64	33.491	32.00	31.40	30.07016	9.0413	48	56
65	33.991	32.50	31.90	30.54001	9.2150	48	57
66	34.491	33.00	32.40	31 00986	9.3887	54	58
67	34.991	33.50	32.90	31.47970	9.5623	54	59
68	35.492	34.00	33.40	31.94955	9.7358	54	60
69	35.992	34.50	33.90	32.41940	9.9092	54	61
70	36.492	35.00	34.40	32.88924	10.0825	54	62
71	36.992	35.50	34.90	33.35909	10.2557	54	63
72	37.492	36.00	35.40	33.82893	10.4289	60	64
73	37.992	36.50	35.90	34.29878	10.6020	60	65
74	38.492	37.00	36.40	34.76863	10.7750	60	66
75	38.993	37.50	36.90	35.23847	10.9480	60	67

Table A-4. (Continued)

INTERNAL GEAR DRIVES

PROPORTIONS OF 1 DP INTERNAL GEARS-20° FULL DEPTH INVOLUTE BASIC RACK MESHING WITH 16 TOOTH PINIONS AND LARGER

No. OF TEETH	ROOT RADIUS	PITCH RADIUS	INTERNAL RADIUS	BASE RADIUS	$\sqrt{R_i^2 - R_b^2}$	LARGEST PINION
N_2	R_r	R	R_i	R_b		N_1
76	39.493	38.00	37.40	35.70832	11.1209	68
77	39.993	38.50	37.90	36.17817	11.2938	69
78	40.493	39.00	38.40	36.64801	11.4666	70
79	40.993	39.50	38.90	37.11786	11.6393	71
80	41.493	40.00	39.40	37.58770	11.8120	72
81	41.993	40.50	39.90	38.05755	11.9846	73
82	42.493	41.00	40.40	38.52740	12.1572	74
83	42.993	41.50	40.90	38.99724	12.3298	75
84	43.493	42.00	41.40	39.46709	12.5023	76
85	43.993	42.50	41.90	39.93694	12.6748	77
86	44.494	43.00	42.40	40.40678	12.8472	78
87	44.994	43.50	42.90	40.87663	13.0196	79
88	45.494	44.00	43.40	41.34648	13.1919	80
89	45.994	44.50	43.90	41.81632	13.3643	81
90	46.494	45.00	44.40	42.28617	13.5366	82
91	46.994	45.50	44.90	42.75601	13.7088	83
92	47.494	46.00	45.40	43.22586	13.8811	84
93	47.994	46.50	45.90	43.69571	14.0532	85
94	48.494	47.00	46.40	44.16555	14.2254	86
95	48.994	47.50	46.90	44.63540	14.3976	87
96	49.494	48.00	47.40	45.10525	14.5697	88
97	49.995	48.50	47.90	45.57509	14.7418	89
98	50.495	49.00	48.40	46.04494	14.9138	90
99	50.995	49.50	48.90	46.51478	15.0859	91
100	51.495	50.00	49.40	46.98463	15.2579	92
101	51.995	50.50	49.90	47.45448	15.4299	93
102	52.495	51.00	50.40	47.92432	15.6019	94
103	52.995	51.50	50.90	48.39417	15.7738	95
104	53.445	52.00	51.40	48.86402	15.9457	96
105	53.995	52.50	51.90	49.33386	16.1176	97
106	54.495	53.00	52.40	49.80371	16.2895	98
107	54.995	53.50	52.90	50.27356	16.4614	99
108	55.495	54.00	53.40	50.74340	16.6333	100
109	55.995	54.50	53.90	51.21325	16.8051	101
110	56.495	55.00	54.40	51.68309	16.9769	102
111	56.995	55.50	54.90	52.15294	17.1487	103
112	57.495	56.00	55.40	52.62279	17.3205	104
113	57.995	56.50	55.90	53.09263	17.4923	105
114	58.495	57.00	56.40	53.56248	17.6641	106
115	58.995	57.50	56.90	54.03233	17.8358	107
116	59.495	58.00	57.40	54.50217	18.0075	108
117	59.996	58.50	57.90	54.97202	18.1793	109
118	60.496	59.00	58.40	55.44186	18.3510	110
119	60.996	59.50	58.90	55.91171	18.5227	111
120	61.496	60.00	59.40	56.38156	18.6943	112
121	61.996	60.50	59.90	56.85140	18.8660	113
122	62.496	61.00	60.40	57.32125	19.0377	114
123	62.996	61.50	60.90	57.79110	19.2093	115
124	63.496	62.00	61.40	58.26094	19.3809	116
125	63.996	62.50	61.90	58.73079	19.5526	117

Appendix D
Applications and Suggested Quality Numbers[1]

[1]Extracted from AGMA GEAR HANDBOOK (AGMA 390-03, 1971) with the permission of the publisher, American Gear Manufacturers Association, 1500 King St., Suite 201, Alexandria, VA 22314.

This section includes a tabulation of many industrial and end use applications for spur, helical, herringbone, bevel & hypoid gearing, racks and fine-pitch worms and worm gearing. A typical AGMA Quality Number range is shown for each of the many industries & applications in this table. When selecting a Quality Number for an industry or an application which is not shown, use a similar industry or application as a guide.

The AGMA Quality Number shown opposite each item of equipment identifies the quality of gearing generally used. There may be certain designs or operating conditions that would justify specifying gears to a lower or higher Quality Number.

Locate the application desired and obtain the AGMA Quality Number suggested, In the interest of economy, use the lower Quality Number shown, unless some of the conditions of the equipment or its operation indicate the use of a higher Quality Number.

Note: Quality Numbers selected from this table should be preceded by the letter "Q" when shown in specifications or on drawings.

Table A-5.

Application	*Quality Numbers	Application	*Quality Numbers
AEROSPACE		BAILING MACHINE	5– 7
Actuators	7–11		
Control Gearing	10–12	BOTTLING INDUSTRY	
Engine Accessories	10–13	Capping	6– 7
Engine Power	10–13	Filling	6– 7
Engine Starting	10–13	Labeling	6– 7
Loading Hoist	7–11	Washer, Sterilizer	6– 7
Propeller Feathering	10–13		
Small Engines	12–13	BREWING INDUSTRY	
		Agitator	6– 8
AGRICULTURE		Barrel Washer	6– 8
Baler	3– 7	Cookers	6– 8
Beet Harvester	5– 7	Filling Machines	6– 8
Combine	5– 7	Mash Tubs	6– 8
Corn Picker	5– 7	Pasteurizer	6– 8
Cotton Picker	5– 7	Racking Machine	6– 8
Farm Elevator	3– 7		
Field Harvester	5– 7	BRICK-MAKING MACHINERY	5– 7
Peanut Harvester	3– 7		
Potato Digger	5– 7	BRIDGE MACHINERY	5– 7
AIR COMPRESSOR	10–11	BRIQUETTE MACHINES	5– 7
AUTOMOTIVE INDUSTRY	10–11		

*Quality Numbers are Inclusive, from lowest to highest numbers shown.

Table A-5. (Continued)

Application	*Quality Numbers
CEMENT INDUSTRY	
(Quarry Operation)	
Conveyor	5– 6
Crusher	5– 6
Diesel-Electric Locomotive	8– 9
Electric Dragline (cast gear)	3
(cut gear)	6– 8
Electric Locomotive	6– 8
Electric Shovel (cast gear)	3
(cut gear)	6– 8
Elevator	5– 6
Locomotive Crane (cast gear)	3
(cut gear)	5– 6
(Plant Operation)	
Air Separator	5– 6
Ball Mill	5– 7
Compeb Mill	5– 6
Conveyor Mill	5– 6
Cooler	5– 6
Elevator	5– 6
Feeder	5– 6
Filter	5– 6
Kiln	5– 6
Kiln Slurry Agitator	5– 6
Overhead Crane	5– 6
Pug, Rod, and Tube Mills	5– 6
Pulverizer	5– 6
Raw and Finish Mill	5– 6
Rotary Dryer	5– 6
Slurry Agitator	5– 6
CHEWING GUM INDUSTRY	
Chicle Grinder	6– 8
Coater	6– 8
Mixer-Kneader	6– 8
Molder-Roller	6– 8
Wrapper	6– 8
CHOCOLATE INDUSTRY	
Glazer, Finisher	6– 8
Mixer, Mill	6– 8
Molder	6– 8
Presser, Refiner	6– 8
Tampering	6– 8
Wrapper	6– 8
CLAY WORKING MACHINERY	5– 7
COMMERCIAL METERS	
Gas	7– 9
Liquid, Water, Milk	7– 9
Parking	7– 9
COMPUTING AND ACCOUNTING MACHINES	
Accounting – Billing	9–10
Adding Machine – Caluclator	7– 9
Addressograph	7
Bookkeeping	9–10
Cash Register	7
Comptometer	6– 8
Computing	10–11
Data Processing	7– 9
Dictating Machine	9
Typewriter	8
CONSTRUCTION EQUIPMENT	
Backhoe	6– 8
Cranes, Open Gearing	3– 6
Enclosed Gearing	6– 8
Ditch Digger	3– 8
Transmission	6– 8
Drag Line	5– 8
Dumpster	6– 8
Paver, Loader	3
Transmission	8
Mixer	3– 5
Swing Gear	3– 5
Mixing Bucket	3
Shaker	8
Shovels, Open Gearing	3– 6
Enclosed Gearing	6– 8
Stationary Mixer, Transmission	8
Drum Gears	3– 5
Stone Crusher, Transmission	8
Conveyor	6
Truck Mixer, Transfer Case	9
Drum Gears	3– 5
CRANES	
Boom Hoist	5– 6
Gantry	5– 6
Load Hoist	5– 7
Overhead	5– 6
Ship	5– 7
CRUSHERS	
Ice, Feed	6– 8
Portable and Stationary	6– 8
Rock, Ore, Coal	6– 8
DAIRY INDUSTRY	
Bottle Washer	6– 7
Homogenizer	7– 9
Separator	7– 9
DAMS AND LOCKS	
Tainter Gates	5– 7
DISH WASHER	
Commercial	5– 7
DISTILLERY INDUSTRY	
Agitator	5– 7
Bottle Filler	5– 7
Conveyor, Elevator	6– 7
Grain Pulverizer	6– 8
Mash Tub	5– 7
Mixer	5– 7
Yeast Tub	5– 7
ELECTRIC FURNACE	
Tilting Gears	5– 7
ELECTRONIC INSTRUMENT CONTROL AND GUIDANCE SYSTEMS	
Accelerometer	10–12
Airborne Temperature Recorder	12–13
Aircraft Instrument	12
Altimeter-Stabilizer	9–11

*Quality Numbers are inclusive, from lowest to highest numbers shown.

Table A-5. (Continued)

Application	*Quality Numbers
Analog Computer	10–12
Antenna Assembly	7– 9
Anti-Aircraft Detector	12
Automatic Pilot	9–11
Digital Computer	10–12
Gun-Data Computer	12–13
Gyro Caging Mechanism	10–12
Gyroscope-Computer	12–13
Pressure Transducer	12–13
Radar, Sonar, Tuner	10–12
Recorder, Telemeter	10–12
Servo System Component	9–11
Sound Detector	9
Transmitter, Receiver	10–12
ENGINES	
Diesel, Semi-Diesel and Internal Combustion	
Engine Accessories	10–12
Supercharger	10–12
Timing Gearings	10–12
Transmission	8–10
FARM EQUIPMENT	
Milking Machine	6– 8
Separator	8–10
Sweeper	4– 6
FLOUR MILL INDUSTRY	
Bleacher	7– 8
Grain Cleaner	7– 8
Grinder	7– 8
Hulling	7– 8
Milling, scouring	7– 8
Polisher	7– 8
Separator	7– 8
FOUNDRY INDUSTRY	
Conveyor	5– 6
Elevator	5– 6
Ladle	5– 6
Molding Machine	5– 6
Overhead Cranes	5– 6
Sand Mixer	5– 6
Sand Slinger	5– 6
Tumbling Mill	5– 6
HOME APPLIANCES	
Blender	6– 8
Mixer	7– 9
Timer	8–10
Washing Machine	8–10
MACHINE TOOL INDUSTRY	
Hand Motion (other than Indexing and Positioning)	6– 9
Feed Drives	8 and up
Speed Drives	8 and up
Multiple Spindel Drives	8 and up
Power Drives, 0-800 FPM	6– 8
800-2000 FPM	8–10
2000-4000 FPM	10–12
Over 4000 FPM	12 and up
Indexing and Positioning – Approximate Positioning	6–10
Accurate Indexing and Positioning	12 and up

Application	*Quality Numbers
MARINE INDUSTRY	
Anchor Hoist	6– 8
Cargo Hoist	7– 8
Conveyor	5– 7
Davit Gearing	5– 7
Elevator	6– 7
Small Propulsion	
Steering Gear	10–12
Winch	8
	5– 8
METAL WORKING	
Bending Roll	5– 7
Draw Bench	6– 8
Forge Press	5– 7
Punch Press	5– 7
Roll Lathe	5– 7
MINING AND PREPARATION	
Agitator	
Breaker	5– 6
Car Dump	5– 6
Car Spotter	5– 7
Centrifugal Drier	7– 8
Clarifier	7– 8
Classifier	7– 8
Coal Digger	6–10
Concentrator	5– 6
Continuous Miner	6– 7
Cutting Machine	6–10
Conveyor	5– 7
Drag Line, Open Gearing	3– 6
Enclosed Gearing	6– 8
Drills	5– 6
Drier	5– 6
Electric Locomotive	6– 8
Elevator	5– 6
Feeder	6– 8
Flotation	5– 6
Grizzly	5– 6
Hoists, Skips	7– 8
Loader (Underground)	5– 8
Rock Drill	5– 6
Rotary Car Dump	6– 8
Screen (Rotary)	7– 8
Screen (Shaking)	7– 8
Separator	5– 6
Sedimentation	5– 6
Shaker	6– 8
Shovel	3– 8
Triple Gearing	5– 7
Washer	6– 8
PAPER AND PULP	
Bag Machines	6– 8
Bleacher, Decker	
Box Machines	6– 8
Building Paper	6– 8
Calender	6– 8
Chipper	6– 8
Coating	6– 8
Digester	
Envelope Machines	6– 8
Food Container	6– 8

* Quality Numbers are inclusive, from lowest to highest numbers shown.

Table A-5. (Continued)

Application	*Quality Numbers
Glazing	6– 8
Grinder	
Log Conveyor – Elevator	5– 7
Mixer, Agitator	6– 8
Paper, Machine	
Auxiliary	8– 9
Main Drive	10–12
Press, Couch, Drier Rolls	6– 8
Save-all	
Slitting	10–12
Steam Drum	6– 8
Varnishing	6– 8
Wall Paper Machines	6– 8
PAVING INDUSTRY	
Aggregate Drier	5– 7
Aggregate Spreader	5– 7
Asphalt Mixer	5– 7
Ashpalt Spreader	5– 7
Concrete Batch Mixer	5– 7
PHOTOGRAPHIC EQUIPMENT	
Aerial	10–12
Commercial	8–10
PRINTING INDUSTRY	
Press, Book	9–11
Flat	9–11
Magazine	9–11
Newspaper	9–11
Roll Reels	6– 7
Rotary	9–11
Book Binding	
PUMP INDUSTRY	
Liquid	10–12
Rotary	6– 8
Slush-duplex-triplex	6– 8
Vacuum	6– 8
QUARRY INDUSTRY	
Conveyor-Elevator	6– 7
Crusher	5– 7
Rotary Screen	7– 8
Shovel-Electric-Diesel	
RADAR AND MISSILE	
Antenna Elevating	8–10
Data Gear	10–12
Launch Pad Azimuth	8
Ring Gear	9–12
Rotating Drive	10–12
RAILROADS	
Construction Hoist	5– 7
Wrecking Crane	6– 8
RUBBER AND PLASTICS	
Boot and Shoe Machines	6– 8
Drier, Press	6– 8
Extruder, Strainer	6– 8
Mixer, Tuber	6– 8
Refiner, Calender	5– 7
Rubber Mill, Scrap Cutter	5– 7
Tire Building	6– 8
Tire Chopper	5– 7
Washer, Banbury Mixer	5– 7
SMALL POWER TOOLS	
Bench Grinder	6– 8
Drills-Saws	7– 9
Hair Clipper	7– 9
Hedge Clipper	7– 9
Sander, Polisher	8–10
Sprayer	6– 8
SPACE NAVIGATION	
Sextant and Star Tracker	13 and up
STEEL INDUSTRY	
Auxiliary and Miscellaneous Drives	
Annealing Furnance	5– 6
Bending Roll	5– 6
Blooming-Mill Manipulator	5– 6
Blooming-Mill Rack and Pinion	5– 6
Blooming-Mill Side Guard	5– 6
Car Haul	5– 6
Coil Conveyor	5– 6
Edger Drives	5– 6
Electrolytic Line	6– 7
Flange-Machine Ingot Buggy	5– 6
Leveler	6– 7
Magazine Pusher	6– 7
Mill Shear Drives	6– 7
Mill Table Drives (under 800 ft/min)	5– 6
Mill Table Drives (over 800-1800 ft/min)	6– 7
Mill Table Drives (over 1800 ft/min)	8
Nail and Spike Machine	5– 6
Piler	5– 6
Plate Mill Rack and Pinion	5– 6
Plate Mill Side Guards	5– 6
Plate Turnover	5– 6
Preheat Furnace Pusher	5– 6
Processor	6– 7
Pusher Rack and Pinion	5– 6
Rotary Furnace	5– 6
Shear Depress Table	5– 6
Slab Squeezer	5– 6
Slab-Squeezer Rack and Pinion	5– 6
Slitter, Side Trimmer	6– 7
Tension Reel	6– 7
Tilt Table, Upcoiler	5– 6
Transfer Car	5– 6
Wire Drawing Machine	6– 7
Blast Furnace, Coke Plant	
Open-Hearth and Soaking Pits	
Miscellaneous Drives	
Bessemer Tilt-Car Dump	5– 6
Coke Pusher, Distributor	5– 6
Conveyor, Door Lift	5– 6
Electric-Furnace Tilt	5– 6
Hot Metal Car Tilt	5– 6
Hot Metal Charger	5– 6
Jib Hoist, Dolomite Machine	5– 6
Larry Car,	5– 6
Mixing Bin, Mixer Tilt	5– 6

*Quality Numbers are inclusive, from lowest to highest numbers shown.

Table A-5. (Continued)

Application	*Quality Numbers
Ore Crusher, Pig Machine	5– 6
Pulverizer, Quench Car	5– 6
Shaker, Stinter Conveyor	5– 6
Stinter Machine Skip Hoist	5– 6
Slag Crusher, Slag Shovel	5– 6
Primary and Secondary Rolling Mill Drives	
Blooming and Plate Mill	5– 6
Heavy Duty Hot Mill Drives	5– 6
Slabbing and Strip Mill	5– 6
Hot Mill Drives	
Sendzimer-Stekel	7– 8
Tandem-Temper-Skin	6– 7
Cold Mill Drives	
Bar, Merchant, Rail, Rod	5– 6
Structural, Tube	5– 6
Mill Gearing	
Billet Mills	
Free Roughing	5– 6
Tandem Roughing	5– 6
Finishing	5– 6
Cold Mills	
Reversing	7– 8
Tandem	7– 8
Temper	7– 8
Foil	7– 8
Hot Mills	
Blooming and Slabbing Mills	5– 6
Continuous Hot Strip Mills	5– 6
Free Reversing Roughing	5– 6
Tandem Roughing (including Scalebreaker)	5– 6
Finishing	6– 7
Merchant Mills	
Roughing	6– 7
Intermediate	7– 8
Finishing	7– 9
Plate Mills	
Reversing Roughing	5– 6
Unidirectional Roughing	5– 6
Unidirectional Finishing	5– 6
Rod Mills	
Roughing	6– 7
Intermediate	7– 8
Finishing	10–12
High Speed	12–14
Skelp Mills	
Roughing	6– 7
Intermediate	7– 8
Finishing	7– 9
Structural and Rail Mills	
Heavy	
Reversing Rougher	5– 6
Finishing	5– 6
Light	
Roughing	5– 6
Finishing	5– 6
Overhead Cranes	
Billet Charger, Cold Mill	5– 6
Bucket Handling	5– 6
Car Repair Shop	5– 6
Cast House, Coil Storage	5– 6
Charging Machine	5– 6
Cinder Yard, Hot Top	5– 6
Coal and Ore Bridges	5– 6
Electric Furnace Charger	5– 6
Hot Metal, Ladle	5– 6
Hot Mill, Ladle House	5– 6
Jib Crane, Motor Room	5– 6
Mold Yard, Rod Mill	5– 6
Ore Unloader, Stripper	5– 6
Overhead Hoist	5– 6
Pickler Building	5– 6
Pig Machine, Sand House	5– 6
Portable Hoist	5– 6
Scale Pit, Shipping	5– 6
Scrap Balers and Shears	5– 6
Scrap Preparation	5– 6
Service Shops	5– 6
Skull Cracker	5– 6
Slab Handling	5– 6
Precision Gear Drives	
Diesel Electric Gearing	8– 9
Flying Shear	9–10
Shear Timing Gears	9–10
High Speed Reels	8– 9
Locomotive Timing Gears	9–10
Pump Gears	8– 9
Tube Reduction Gearing	8– 9
Turbine	9–10
MISCELLANEOUS	
Clocks	6
Counters	7– 9
Fishing Reel	6
Gauges	8–10
IBM Card Puncher, Sorter	8
Metering Pumps	7– 8
Motion Picture Equipment	8
Popcorn Machine, Commercial	6– 7
Pumps	5– 7
Sewing Machine	8
Slicer	7– 8
Vending Machines	6– 7

*Quality Numbers are inclusive, from lowest to highest numbers shown.

Appendix E
Gear Geometry

We list, alphabetically, the most commonly used terms in gearing. Following each term, where applicable, is the symbol for use in gear engineering formulas and equations, in parentheses.

Addendum (a)

The addendum of a gear tooth is the radial distance from the nominal pitch circle to the top of the tooth.

Addendum—Chordal (a_c)

The *chordal addendum* is used for the purpose of setting a gear tooth vernier to measure the tooth thickness, which is the chord length subtended by the two flanks of a gear tooth at the nominal pitch circle. This must be calculated from the known circular tooth thickness. The distance from the chord to the top of the tooth is known as the *chordal addendum.*

Addendum Angle (α)

In a bevel gear, this is the angle between elements of the pitch cone and the face cone.

Angular Pitch (φ_N)

This is the angle subtended by the circular pitch, usually expressed in radians. Thus $\varphi_N = \frac{2\pi}{N}$ radians (or $\frac{360°}{N}$ degrees).

Apex to Back

In a bevel gear, this is the distance from the apex of the pitch cone to a locating surface at the back of the gear.

Arc of Action (Q_t)

This is the arc on the pitch circle through which a tooth profile moves from the beginning to the end of contact with a mating tooth profile.

Arc of Approach (Q_a)

This is the arc on the pitch circle through which a tooth profile moves from its beginning of contact with a mating tooth profile until it reaches the pitch point.

Arc of Recess (Q_r)

This is the arc on the pitch circle through which a tooth profile moves from contact with a mating tooth profile at the pitch point until contact ends.

Axial Pitch (P_x)

This is the circular pitch in the axial direction in helical gears. It is, therefore, normal CP divided by the sine of the helix angle.

Axial Runout

Also known as "wobble," this is the runout of the gear in the axial direction, measured at just below the root circle. It is expressed as "total indicator reading."

Back Angle

In a bevel gear, this is the angle between an element of the back cone and the plane of rotation. It is usually equal to the pitch angle.

Back Cone

The elements of the back cone of a bevel gear extend from the outside diameter of the gear blank to its axis and are perpendicular to the elements of the pitch cone.

Back Cone Distance

This is the distance along an element of the back cone from its apex to the pitch circle.

Backlash

This is the gap between the profiles of meshing gear teeth when the driving side of a tooth is held tight against the driven at operating center distance. It can be measured with a feeler gauge.

ALL types of gears must have backlash. See Chapter 6.

Base Circle

The base circle is the circle from which involute gear tooth profiles are developed.

Base Diameter (D_b)

This is the diameter of the base circle of an involute gear.

Base Diameter—Cone® Gear

This is the base circle diameter of a Cone® double-enveloping worm gear at the center of the tooth face.

Base Pitch (P_b)

This is the pitch, in inches or millimeters, measured along the circumference of the base circle or along the *line of action*. It is, therefore, the circumference of the base circle divided by the number of teeth. It is the same if measured along the line of action because the corresponding profiles of involute gear teeth are parallel curves, and the base pitch is the constant distance between them along the common normal in the direction of rotation, which, by definition, is the line of action.

Bottom Land

This is the surface at the bottom of a tooth space which adjoins the fillets. For full fillet teeth there is no bottom land as such.

Center Distance (C)

This is the distance between the centers of two gears in mesh with each other on nonintersecting shafts. It is measured along the mutual perpendicular to the shafts, called the *line of centers*. It applies to spur gears, parallel-axis or crossed-axis helical gears, double-helical and herringbone gears, and worm gearing.

Chart, Inspection

An inspection chart is the generated recording, or trace, from an inspection machine used to display a measured variation of gear geometry.

Chordal Thickness (t_c)

This is the length of the chord subtended by the two flanks of a gear tooth, usually at the nominal pitch circle diameter.

Circular Pitch (p)

This is the distance between the corresponding profiles of two adjacent teeth as measured along the pitch circle. It is, therefore, the circumference of the pitch circle divided by the number of teeth.

Circular Tooth Thickness (t)

This is the length of arc between the two flanks of a gear tooth, usually at the nominal pitch circle diameter.

Clearance (c)

This is the distance between the root circle of a gear and the outside, or addendum, circle of its mate.

Composite Action

This is the variation in center distance when two gears are rolled in tight mesh during a composite action test.

Cone Distance

In a bevel gear, the distance from the base of its cone to the cone's apex is called the *cone distance*.

Conjugate

A term used to describe gear tooth forms which properly mate with each other—they are *conjugate* to each other. From Webster's New World Dictionary, 2nd College Edition, 1986, a definition of "conjugate" is: "reciprocally related and interchangeable as to properties, as two points, lines, etc."

Contact Ratio (m_c)

This is the length of the *arc of action* divided by the *base pitch*. See Chapter 3.

Crowned Teeth

Crowned teeth are often specified for spur, helical, and straight-tooth bevel gears. They are thicker at or near the center of the tooth face than at the ends of the teeth. Crowning is accomplished by crown hobbing, crown shaving, or crown grinding. On straight bevels the trade designation of crowned teeth is Coniflex®, a Gleason Works trademark. The purpose of crowning is to minimize the chance of end bearing of the teeth when there is misalignment in assembly, distortion from heat treating, or deflection under load.

The amount of crown is commonly in the neighborhood of .0003″ per inch of face (.0006″ change in tooth thickness per inch of face). It can be varied but if assembly conditions call for more than this, the designer must be concerned with the compressibility of the tooth surfaces and tooth deflection. If the crown exceeds the amount the teeth will compress or deflect under full load to provide adequate tooth bearing area, the full face of the gearing cannot be used in determining the load rating.

Dedendum (d)

The dedendum of a gear tooth is the radial distance from the nominal pitch circle to its root circle.

Dedendum Angle (δ)

In a bevel gear, this is the angle between elements of the root cone and pitch cone.

Diametral Pitch (Transverse) (P_d or DP)

In spur gearing and transversal pitch helical gearing, this is the number of teeth per inch of pitch diameter (*D* or *PD*). Thus: $P_d = N/D$, or $DP = N/PD$.

It is seen that as *DP* decreases the tooth size increases. For example 1 *DP* teeth are larger than 2 *DP* teeth.

The relationship between *DP* and *CP* is: $DP = 3.1416/CP$, or $CP = 3.1416/DP$.

Eccentricity

See "Runout."

Face Width (F)

This is the length of the teeth of spur gearing, or in helical or herringbone gearing it is the length of the teeth times the cosine of the helix angle.

Fillet Radius (r_f)

This is the radius of an arc approximating the root fillet curve.

Form Diameter

This is the diameter of the circle that intersects the trochoid formed by the cutting tool and the involute curve tooth profile. Tooth contact cannot take place below the form diameter.

Full Depth

This term is used to describe gear teeth whose working depth equals 2.0 divided by the diametral pitch.

Gear Ratio (m_G)

Also known as *speed ratio*, this is the number of teeth in the driven member (usually the larger, or the gear) divided by the number of teeth in the driver (usually the smaller, or the pinion).

Heel

This is what the thickest end of a bevel gear tooth is called. See "Bevel Gears—Straight" in Chapter 1.

Helix Angle (ψ)

In helical and herringbone gears, this is the angle between the gear teeth and the axis of rotation at the pitch circle.

Indicating Over Pins

This is the measurement of the change in radial distance over pins or wires placed in the tooth space to determine runout with reference to the axis of rotation.

Index Variation (V_x)

This is the displacement of any tooth form from its theoretical position relative to a datum tooth. Measurements are usually linear, near the middle of the functional tooth profile, and if made normal to the tooth surface, should be corrected to the transverse plane. Distinction is made as to the algebraic sign of this reading.

Involute Curve

The curve of the profile of most gear teeth. See Chapter 3.

Lead (L)

This is the axial advance of a helix for one complete turn, as in the threads of cylindrical worms and the teeth of helical gears.

Lead Angle (λ)

The angle between any helix and a plane of rotation. It is the complement of the helix angle and is used for convenience in worms and hobs. It is understood to be at the standard pitch diameter.

Length of Action (Z)

This is the distance along the *line of action* a tooth moves from the beginning to the end of contact with its mating tooth.

Line of Action

This is the straight line that is tangent to the base circles of two mating gears. Its intersection with the centerline of the two base circles defines the *pitch point*. The line of action is the path the teeth follow while in contact.

Long-and-Short Addendum

This term is used to describe modified gear teeth in which the pinion teeth are cut on an oversize blank and the gear teeth are cut on a blank that is undersize by the same amount. This design is usually specified when an oversize pinion is necessary to avoid *undercut* and standard center distance must be maintained.

Master Gear

A master gear is used in checking a production gear in a composite action inspection test, by rolling the master with the production gear in tight mesh (spring loaded) and measuring the variation in centers. Usually the master gear is hardened, with ground teeth, and produced to a high degree of accuracy.

Measurement over Pins

It is common practice to check spur gears by taking measurements of diameters over pins, which are not pins at all, but are short sections of hardened and ground steel rods of various diameters. To measure a gear a pin is laid in tooth spaces opposite each other and diameter measurements are made over the tops of the pins. If the gear has an even number of teeth the pins are placed 180° apart, if an odd number the pins are placed as close as possible to 180° apart. The diameter of the pin to be used is selected to suit the pitch of the gear to be checked. It must be of such size that it will rest in the tooth space, touching the tooth flanks at or slightly above the pitch line, stick out above the OD of the gear, and must not touch the root diameter. Standard pin sizes have been established, and gear makers and users can buy standard size pins for the commonly used pitches of teeth. There are published tables to enable determination of measurements over pins for the standard pitches without going through lengthy calculations.

Helical gears can also be checked over pins, although theoretically balls should be used instead of pins because a helical tooth is curved. The calculations should be made for ball measurement, but most gear manufacturers then use pins for the actual measurement. There is a slight difference between the calculated ball measurement and the actual pin measurement, the difference increasing as the helix angle increases, but is usually so slight that it can be ignored.

Module (m)

This is the ratio of the pitch diameter in millimeters to the number of teeth.

Module (Normal) (m_n)

This is the value of m in the normal plane of a helical gear or worm. Thus: $m_n = m \cos \psi$.

Mounting Distance

In bevel gearing, this is the distance from the crossing point of the axes of two bevel gears in mesh with each other to a locating surface on each. It is used in assembling bevel gearing.

Number of Teeth or Threads (N)

This is the number of teeth or threads contained in the whole circumference of the pitch circle.

Pi (π)

This is the circumference of a circle divided by its diameter. It is an irrational number. Its value is 3.14159265 . . . and on and on. The value 3.1416 is used in most engineering calculations, although for calculations requiring accuracy to more than four places, it is common practice to use as many places as shown for the first value given.

To determine the value of Pi mathematically one must start with determination of the circumference by assuming that a polygon of infinite number of sides is inscribed within a circle with a radius of unity. This is a problem in the calculus.

For the reader interested in learning more about Pi, read *A History of Pi*, by Petr Beckman, St. Martin's Press, New York.

Pitch

This is a measure of tooth spacing and size. See "Circular Pitch" and "Diametral Pitch."

Pitch Circle

Envision two discs in contact with each other, one driving the other by friction without slipping. Relative motion would be transmitted from the shaft of one to the shaft of the other in inverse proportion to the diameters of the two discs (if the driven disc is twice the size of the driver, its speed will be half the speed of the driver). If these two discs are provided with gear teeth, the outside diameters of which would be larger and the root diameters of which would be smaller than the diameters of the discs, the circumference of each disc is called the *pitch circle* of the driver and the driven gears.

Pitch Diameter (D) or *(PD)*

Pitch diameter, nominal, is the diameter of the pitch circle of a gear. This is the *PD* for the tooth cutting operation. The nominal pitch circles of two gears in mesh at standard centers will be tangent to each other. The operating pitch circles of two gears when meshing at greater than standard centers, as when the pinion is oversize, will be greater than the nominal pitch circles and now they will be tangent to each other.

Pitch Line

In a rack, this corresponds to the pitch circle of a gear.

Pitch Point

This is the point of tangency of two pitch circles, or of a pitch circle and pitch line, and is on the line of centers.

Point of Contact

This is any point at which two tooth profiles touch each other.

Profile

This is the curve defined by the intersection of a tooth surface and a plane normal to that surface.

Quality Number (Q)

This is the number used to define the quality level of a gear. Complete definitions are provided in Standard ANSI/AGMA 2000-A88 and in somewhat simpler form in AGMA Gear Handbook 390.03.

Rack

A rack is a gear with teeth spaced along a straight line instead of a circle. Its use is in mechanisms requiring conversion of rotary motion (the pinion) into linear motion (the rack). See "Basic Rack," Chapter 1.

Radius of Curvature—Profile

This is the radius of curvature of the profile of a gear tooth, usually at the pitch point or a point of contact.

Runout

A measure of eccentricity relative to the axis of rotation. It is measured in the radial direction and the amount of runout is the difference between the highest and the lowest reading in 360°. For gear teeth runout is usually checked by placing a pin in tooth spaces and rolling past a dial indicator or by rolling with a master gear.

*Shaft Angle (*Ε*)*

The angle between the axes of two nonparallel gear shafts. In a pair of crossed-axis helical gears the shaft angle lies between the oppositely rotating portions of two shafts, as is also true with worm gearing. In bevel gears the shaft angle is the sum of the two pitch angles. In hypoid gears, the shaft angle is given when starting a design and it does not have a fixed relation with the pitch and spiral angles.

Spacing

The term "spacing" is used as a general term to describe the accuracy with which teeth are spaced around the gear.

Span Measurement

This is the measurement of the distance across several teeth of gears too large to use pin measurements. The measurement is made along a line tangent to the base circle. It is used to determine tooth thickness and tooth spacing accuracy.

For such measurements a *span measuring tool* is used, set to touch the flanks of teeth at the ends of each span at or near the middle of the tooth height. A span measuring tool will usually be set by a vernier and will be equipped with a dial indicator to indicate any deviation from the theoretical chord length.

Spiral Angle (ψ)

This is the angle between the tooth and an element of the pitch cone in a spiral bevel gear. It is usually understood to be at the mean cone distance.

Stub Teeth

Stub teeth were developed years ago by Fellows Corporation of Springfield, VT to alleviate the problem of getting rid of undercut in spur pinions with relatively few numbers of teeth.

There are two stub tooth systems, both with 20° PA. The first was the Fellows Stub, designated as a fraction, such as 4/5 *DP*, which says that the pitch of the gear is 4 *DP* with a 5 *DP* tooth depth. A later system is the *AGMA 20° Stub* system. In this system the addendum is .8/*DP* and the dedendum is 1/*DP*. The AGMA system is interchangeable with the Fellows system and has largely replaced it.

Test Radius (R_r)

This is the number used as the test distance between a master and a work gear for a composite action test. The test radius of the master plus the test radius of the work gear is the setup center distance of a composite action test device.

Throat Diameter (d_t)

This is the diameter of the addendum circle at the center of the face of a cylindrical or double-enveloping worm gear.

Throat Form Radius (r_t)

This is the radius of the throat of a cylindrical or double-enveloping worm gear in an axial plane. It is normally generated by the worm gear hob, which is a topping hob, and will be slightly larger than the minor radius of the mating worm.

Tip Radius (r_T)

This is the radius between the outside and side cutting edges of hobs and other gear cutting tools.

Tip Relief

This is an arbitrary modification of the tooth profile near the tip of the teeth to minimize or eliminate tip interference. It is considered desirable for the involute to be a few thousandths minus at the tip—never plus.

Toe

In a bevel gear or pinion this is the thinnest end of a tooth. See "Bevel Gears—Straight" in Chapter 1 for fuller explanation.

Tolerance

This is the amount by which a specific dimension is permitted to vary. It is usually expressed as the difference between the maximum and minimum limits allowed.

Top Land

This is the width, or thickness, of a gear tooth measured at its maximum diameter, in the direction of rotation.

Undercut

For pinions with small numbers of teeth, the cutting tool will cut away that portion of the involute that is near and below the base circle. This cut away portion is called *undercut*, and it increases as the number of teeth becomes less. See Chapter 5.

Whole Depth (h_t)

This is the distance from the top land of a gear tooth to its root.

Working Depth (h_k)

This is the depth at which gear teeth are engaging.

Zone of Action

This is the contact zone for spur or helical gears. It is the rectangular area in the plane of action bounded by the length of action and the face width.

Much of the material in pp. 183 to 196 was taken (some quoted, some paraphrased) from AGMA Standard 1012-F90, "Gear Nomenclature, Definitions of Terms with Symbols," with the permission of the publisher, American Gear Manufacturers Association, 1500 King St., Suite 201, Alexandria, VA 22314.

Appendix F
Glossary of Terms and Symbols
(Listed in Approximate Order of Appearance)

Terms

T	number of teeth
F	face width, inches
PD	pitch diameter, inches (T/DP)
DP	diametral pitch (PD/T)
CP	circular pitch, inches (3.1416/DP)
M	module, millimeters (25.400/DP)
R_b	radius of base circle, inches
r	pitch radius, inches (Figure 3.1a)
R	pitch radius, inches (Figure 3.1b)
ϕ (phi)	pressure angle, degrees
θ (theta)	vector angle, degrees (angle from start of involute curve to pitch point)
P	pitch point of spurs
R_O	outside radius, inches
OD	outside diameter, inches
CD	center distance, inches
TT	tooth thickness, inches
BHN	Brinell Hardness Number
RC	Rockwell Hardness Number (C scale)
RA	Rockwell Hardness Number (A scale)
R30-N	Rockwell Hardness Number (30-N scale)
R15-N	Rockwell Hardness Number (15-N scale)
TS	tensile strength, psi
YP	yield point, psi
R of *A*	reduction of area, %
*E*1	elongation, %
Pi (π)	3.1416: Circumference of a circle divided by its diameter, an irrational number. Determination of the circumference can be accomplished mathematically by assuming that a polygon of infinite number of sides is inscribed within a circle with a radius of unity, which is a calculus problem. The resulting value of Pi is 3.14159265 . . . and on and on. For most engineering calculations the value 3.1416 is used. For the reader interested in learning more about Pi, read *A History of Pi*, by Peter Beckman, St. Martin's Press, New York.
P	axial pitch of a worm, inches
L	lead of a worm, inches ($P \times$ no. of starts)
γ (gamma)	lead angle of a worm, degrees
ϕ_n	normal pressure angle, degrees
V	velocity, rpm
V_s	sliding velocity of a worm, ft/min
n (or N)	speed, rpm
W_t	tangential tooth load, lb

f	coefficient of friction
E	efficiency
HP	horsepower, $= 2 \times \pi \times N \times T/33{,}000$
T	torque, inch-lb
s_c	compressive stress, psi
E	modulus of elasticity, stress/strain (30,000,000 for steel)
M_g	speed ratio
s_{ac}	allowable compressive stress, psi
s_t	tensil stress, psi
Y	tooth form factor

Most of the above terms are what AGMA calls "abbreviations" in AGMA Standard ANSI/AGMA 1012-F90, which lists symbols for use in gear engineering formulas. The AGMA list of symbols is reproduced herein on pp. 201 and 202. Example: For circular pitch the abbreviation is *CP*, the symbol is *p*.

Symbols

Symbols	Terms	Symbols	Terms
A	cone distance	C	center distance
A_i	inner cone distance	c	clearance
A_m	mean cone distance	D	reference standard pitch diameter
A_o	outer cone distance	D_b	base diameter (of an involute gear)
a	addendum	D_c	datum circle
a_c	chordal addendum	D_i	internal diameter
a_{nc}	normal chordal addendum	D_o	outside diameter
B	backlash	D_R	root diameter
b	dedendum	d_t	throat diameter
d_p	operating pitch diameter	r_t	throat-form radius
E	offset of axes	r_T	tip or edge radius of tool
F	face width	t	circular tooth thickness
F_e	effective or active face width	t_b	base circular thickness
F_t	total face width	t_c	chordal thickness
h_k	working depth	t_n	normal circular thickness
h_t	whole depth (tooth depth)	t_{nc}	normal chordal thickness
L	lead	t_T	tooth thickness tolerance
m	module	t_t	transverse circular thickness
m_c	contact ratio	t_x	axial thickness
m_F	face contact ratio	V_{ap}	total accumulated pitch variation
m_G	gear ratio	V_{apk}	total accumulated pitch variation, within a sector of k pitches
m_n	normal module		
m_o	modified contact ratio	V_{cq}	total composite variation (double flank)
m_p	transverse contact ratio		
m_t	total contact ratio	V_{cqT}	total composite tolerance (double flank)
N	numbers of teeth or threads	V_p	pitch variation
N_e	equivalent number of teeth	V_{pA}	allowable pitch variation
P_d	diametral pitch (transverse)	V_{pn}	normal pitch variation
P_{nd}	normal diametral pitch	V_q	tooth-to-tooth composite variation (double flank)
P	circular pitch		
p_b	base pitch	V_{qT}	tooth-to-tooth composite tolerance (double flank)
p_m	true position pitch		
p_n	normal circular pitch	V_r	radial runout
p_N	normal base pitch	V_{rT}	radial runout tolerance
p_t	transverse circular pitch	V_s	spacing variation
p_x	axial pitch	V_x	index variation
p_X	axial base pitch	V_ϕ	profile variation
Q	Quality Number	$V_{\phi T}$	profile tolerance

Symbols	Terms	Symbols	Terms
τ_o	face tip angle	ϕ	pressure angle
δ	dedendum angle	ϕ_n	normal pressure angle
Σ	shaft angle	ϕ_t	transverse pressure angle
ε	involute roll angle	ϕ_x	axial pressure angle
Θ	involute polar angle	ψ	helix angle, spiral angle
Θ_N	angular pitch	ψ_b	base helix angle
λ	lead angle	ψ_i	spiral angle at inner cone distance
λ_b	base lead angle	ψ_m	spiral angle at mean cone distance
λ_o	outside lead angle	ψ_o	outside helix, outer spiral angle
λ_p	pitch lead angle	ψ_p	pitch helix angle
ρ	profile radius of curvature	V_ψ	tooth alignment variation
Q_a	arc of approach	$V_{\psi T}$	tooth alignment tolerance
Q_F	face advance	Z	length of action
Q_r	arc of recess	α	addendum angle
Q_t	arc of action	τ	pitch angle
R_r	test radius	τ_R	root angle
r_f	fillet radius, when constant		

Table on pp. 199–200 are taken from AGMA Standard 1012-F90, with the permission of the publisher, American Gear Manufacturers Association, 1500 King St., Suite 201, Alexandria, VA 22314.

Appendix G
Involute Functions

INVOLUTE FUNCTIONS

12°

Φ	INV Φ	DIFF.	Φ	INV Φ	DIFF.
.00	.00311 705	789	.50	.00352 851	858
.01	2 494	791	.51	3 709	860
.02	3 285	792	.52	4 569	861
.03	4 077	793	.53	5 430	863
.04	4 870	795	.54	6 293	864
.05	5 665	796	.55	7 157	866
.06	6 461	797	.56	8 023	867
.07	7 258	799	.57	8 890	869
.08	8 057	800	.58	.00359 759	869
.09	8 857	801	.59	.00360 628	872
.10	.00319 658	803	.60	1 500	872
.11	.00320 461	805	.61	2 372	875
.12	1 266	805	.62	3 247	875
.13	2 071	807	.63	4 122	877
.14	2 878	808	.64	4 999	879
.15	3 686	810	.65	5 878	880
.16	4 496	811	.66	6 758	881
.17	5 307	812	.67	7 639	883
.18	6 119	814	.68	8 522	884
.19	6 933	815	.69	.00369 406	886
.20	7 748	817	.70	.00370 292	887
.21	8 565	818	.71	1 179	889
.22	.00329 383	819	.72	2 068	890
.23	.00330 202	821	.73	2 958	891
.24	1 023	822	.74	3 849	893
.25	1 845	824	.75	4 742	894
.26	2 669	824	.76	5 636	896
.27	3 493	827	.77	6 532	897
.28	4 320	827	.78	7 429	899
.29	5 147	829	.79	8 328	900
.30	5 976	831	.80	.00379 228	902
.31	6 807	831	.81	.00380 130	903
.32	7 638	834	.82	1 033	904
.33	8 472	834	.83	1 937	906
.34	.00339 306	836	.84	2 843	908
.35	.00340 142	838	.85	3 751	909
.36	0 980	838	.86	4 660	910
.37	1 818	841	.87	5 570	912
.38	2 659	841	.88	6 482	913
.39	3 500	843	.89	7 395	915
.40	4 343	845	.90	8 310	916
.41	5 188	845	.91	.00389 226	918
.42	6 033	848	.92	.00390 144	919
.43	6 881	848	.93	1 063	921
.44	7 729	850	.94	1 984	922
.45	8 579	852	.95	2 906	924
.46	.00349 431	853	.96	3 830	925
.47	.00350 284	854	.97	4 755	926
.48	1 138	856	.98	5 681	928
.49	1 994	857	.99	6 609	930
.50	.00352 851		1.00	.00397 539	

13°

Φ	INV Φ	DIFF.	Φ	INV Φ	DIFF.
.00	.00397 539	931	.50	00445 931	1007
.01	8 470	932	.51	6 938	008
.02	.00399 402	934	.52	7 946	010
.03	.00400 336	936	.53	8 956	011
.04	1 272	937	.54	.00449 967	013
.05	2 209	938	.55	.00450 980	015
.06	3 147	940	.56	1 995	016
.07	4 087	941	.57	3 011	017
.08	5 028	943	.58	4 028	020
.09	5 971	945	.59	5 048	020
.10	6 916	946	.60	6 068	022
.11	7 862	947	.61	7 090	024
.12	8 809	949	.62	8 114	026
.13	.00409 758	950	.63	.00459 140	027
.14	.00410 708	952	.64	.00460 167	028
.15	1 660	954	.65	1 195	030
.16	2 614	954	.66	2 225	032
.17	3 568	957	.67	3 257	033
.18	4 525	958	.68	4 290	035
.19	5 483	959	.69	5 325	037
.20	6 442	961	.70	6 362	038
.21	7 403	962	.71	7 400	039
.22	8 365	964	.72	8 439	041
.23	.00419 329	966	.73	.00469 480	043
.24	.00420 295	967	.74	.00470 523	044
.25	1 262	968	.75	1 567	046
.26	2 230	970	.76	2 613	047
.27	3 200	972	.77	3 660	049
.28	4 172	973	.78	4 709	051
.29	5 145	974	.79	5 760	052
.30	6 119	976	.80	6 812	054
.31	7 095	978	.81	7 866	055
.32	8 073	979	.82	8 921	057
.33	.00429 052	981	.83	.00479 978	059
.34	.00430 033	982	.84	.00481 037	060
.35	1 015	983	.85	2 097	062
.36	1 998	986	.86	3 159	063
.37	2 984	986	.87	4 222	065
.38	3 970	989	.88	5 287	066
.39	4 959	989	.89	6 353	068
.40	5 948	992	.90	7 421	070
.41	6 940	993	.91	8 491	071
.42	7 933	994	.92	.00489 562	073
.43	8 927	996	.93	.00490 635	075
.44	.00439 923	997	.94	1 710	076
.45	.00440 920	1000	.95	2 786	078
.46	1 920	000	.96	3 864	079
.47	2 920	002	.97	4 943	081
.48	3 922	004	.98	6 024	083
.49	4 926	1005	.99	7 107	1084
.50	.00445 931		1.00	.00498 191	

INVOLUTE FUNCTIONS

14°			14°			15°			15°		
φ	INV φ	DIFF.	φ	INV φ	DIFF.	φ	INV φ	DIFF.	φ	INV φ	DIFF.
.00	.00498 191	1 086	.50	.00554 484	1 168	.00	.00614 980	1 254	.50	.00679 851	1 343
.01	.00499 277	087	.51	5 652	170	.01	6 234	256	.51	.00681 194	345
.02	00500 364	089	.52	6 822	172	.02	7 490	258	.52	2 539	347
.03	1 453	091	.53	7 994	173	.03	.00618 748	259	.53	3 886	349
.04	2 544	092	.54	.00559 167	175	.04	.00620 007	261	.54	5 235	350
.05	3 636	094	.55	.00560 342	177	.05	1 268	263	.55	6 585	353
.06	4 730	095	.56	1 519	178	.06	2 531	264	.56	7 938	354
.07	5 825	097	.57	2 697	180	.07	3 795	266	.57	.00689 292	356
.08	6 922	1 099	.58	3 877	182	.08	5 061	268	.58	.00690 648	358
.09	8 021	1 100	.59	5 059	183	.09	6 329	270	.59	2 006	359
.10	.00509 121	102	.60	6 242	185	.10	7 599	272	.60	3 365	362
.11	.00510 223	104	.61	7 427	187	.11	.00628 871	273	.61	4 727	363
.12	1 327	106	.62	8 614	188	.12	.00630 144	275	.62	6 090	365
.13	2 433	106	.63	.00569 802	190	.13	1 419	277	.63	7 455	367
.14	3 539	109	.64	.00570 992	192	.14	2 696	278	.64	.00698 822	369
.15	4 648	110	.65	2 184	194	.15	3 974	281	.65	.00700 191	371
.16	5 758	112	.66	3 378	195	.16	5 255	282	.66	1 562	372
.17	6 870	113	.67	4 573	197	.17	6 537	284	.67	2 934	375
.18	7 983	115	.68	5 770	1 199	.18	7 821	286	.68	4 309	376
.19	.00519 098	117	.69	6 969	1 200	.19	.00639 107	287	.69	5 685	378
.20	.00520 215	118	.70	8 169	202	.20	.00640 394	289	.70	7 063	380
.21	1 333	120	.71	.00579 371	204	.21	1 683	291	.71	8 443	382
.22	2 453	122	.72	.00580 575	205	.22	2 974	293	.72	.00709 825	383
.23	3 575	123	.73	1 780	207	.23	4 267	295	.73	.00711 208	386
.24	4 698	125	.74	2 987	209	.24	5 562	296	.74	2 594	387
.25	5 823	126	.75	4 196	211	.25	6 858	1 298	.75	3 981	389
.26	6 949	129	.76	5 407	212	.26	8 156	1 300	.76	5 370	391
.27	8 078	129	.77	6 619	214	.27	.00649 456	302	.77	6 761	393
.28	.00529 207	132	.78	7 833	216	.28	.00650 758	303	.78	8 154	395
.29	00530 339	133	.79	.00589 049	218	.29	2 061	306	.79	.00719 549	397
.30	1 472	135	.80	.00590 267	219	.30	3 367	307	.80	.00720 946	1 398
.31	2 607	136	.81	1 486	221	.31	4 674	309	.81	2 344	1 400
.32	3 743	138	.82	2 707	223	.32	5 983	310	.82	3 744	403
.33	4 881	140	.83	3 930	224	.33	7 293	313	.83	5 147	404
.34	6 021	142	.84	5 154	226	.34	8 606	314	.84	6 551	406
.35	7 163	143	.85	6 380	228	.35	.00659 920	316	.85	7 957	407
.36	8 306	145	.86	7 608	230	.36	.00661 236	318	.86	.00729 364	410
.37	.00539 451	146	.87	.00598 838	232	.37	2 554	320	.87	00730 774	412
.38	00540 597	148	.88	.00600 070	232	.38	3 874	321	.88	2 186	413
.39	1 745	150	.89	1 302	235	.39	5 195	324	.89	3 599	415
.40	2 895	151	.90	2 537	236	.40	6 519	325	.90	5 014	417
.41	4 046	153	.91	3 773	238	.41	7 844	327	.91	6 431	419
.42	5 199	155	.92	5 011	241	.42	.00669 171	328	.92	7 850	421
.43	6 354	157	.93	6 252	241	.43	.00670 499	331	.93	.00739 271	423
.44	7 511	158	.94	7 493	244	.44	1 830	332	.94	.00740 694	425
.45	8 669	159	.95	8 737	245	.45	3 162	334	.95	2 119	426
.46	.00549 828	162	.96	.00609 982	247	.46	4 496	336	.96	3 545	429
.47	.00550 990	163	.97	.00611 229	249	.47	5 832	338	.97	4 974	430
.48	2 153	165	.98	2 478	250	.48	7 170	340	.98	6 404	432
.49	3 318	1 166	.99	3 728	1 252	.49	8 510	1 341	.99	7 836	1 435
.50	.00554 484		1.00	.00614 980		.50	.00679 851		1.00	.00749 271	

INVOLUTE FUNCTIONS

16° Φ	INV Φ	DIFF.	16° Φ	INV Φ	DIFF.	17° Φ	INV Φ	DIFF.	17° Φ	INV Φ	DIFF.
.00	.00749 271	1 436	.50	.00823 417	1 532	00	.00902 471	1 632	.50	.00986 617	1 736
.01	.00750 707	437	.51	4 949	535	01	4 103	635	.51	.00988 353	738
.02	2 144	440	.52	6 484	536	02	5 738	636	.52	.00990 091	741
.03	3 584	442	.53	8 020	538	03	7 374	639	.53	1 832	742
.04	5 026	444	.54	.00829 558	540	04	.00909 013	640	.54	3 574	745
.05	6 470	445	.55	.00831 098	543	.05	.00910 653	643	.55	5 319	747
.06	7 915	447	.56	2 641	544	.06	2 296	645	.56	7 066	748
.07	.00759 362	450	.57	4 185	546	.07	3 941	646	.57	.00998 814	751
.08	.00760 812	451	.58	5 731	548	.08	5 587	649	.58	.01000 565	754
.09	2 263	453	.59	7 279	550	.09	7 236	651	.59	2 319	755
.10	3 716	455	.60	.00838 829	552	.10	.00918 887	653	.60	4 074	757
.11	5 171	457	.61	.00840 381	554	.11	.00920 540	655	.61	5 831	760
.12	6 628	459	.62	1 935	556	.12	2 195	656	.62	7 591	761
.13	8 087	460	.63	3 491	558	.13	3 851	660	.63	.01009 352	764
.14	.00769 547	463	.64	5 049	560	.14	5 511	661	.64	.01011 116	766
.15	.00771 010	465	.65	6 609	562	.15	7 172	663	.65	2 882	768
.16	2 475	466	.66	8 171	564	.16	.00928 835	665	.66	4 650	770
.17	3 941	468	.67	.00849 735	566	.17	.00930 500	667	.67	6 420	772
.18	5 409	471	.68	.00851 301	568	.18	2 167	669	.68	8 192	775
.19	6 880	472	.69	2 869	570	.19	3 836	672	.69	.01019 967	776
.20	8 352	474	.70	4 439	572	.20	5 508	673	.70	.01021 743	779
.21	.00779 826	476	.71	6 011	574	.21	7 181	676	.71	3 522	781
.22	.00781 302	478	.72	7 585	576	.22	.00938 857	677	.72	5 303	783
.23	2 780	480	.73	.00859 161	578	.23	.00940 534	680	.73	7 086	785
.24	4 260	482	.74	.00860 739	580	.24	2 214	682	.74	.01028 871	787
.25	5 742	483	.75	2 319	582	.25	3 896	684	.75	.01030 658	790
.26	7 225	486	.76	3 901	584	.26	5 580	685	.76	2 448	791
.27	.00788 711	488	.77	5 485	586	.27	7 265	688	.77	4 239	794
.28	.00790 199	489	.78	7 071	588	.28	.00948 953	691	.78	6 033	796
.29	1 688	492	.79	.00868 659	590	.29	.00950 644	692	.79	7 829	1 798
.30	3 180	493	.80	.00870 249	592	.30	2 336	694	.80	.01039 627	1 800
.31	4 673	495	.81	1 841	594	.31	4 030	696	.81	.01041 427	802
.32	6 168	498	.82	3 435	595	.32	5 726	1 698	.82	3 229	805
.33	7 666	1 499	.83	5 030	1 598	.33	7 424	1 701	.83	5 034	807
.34	.00799 165	1 501	.84	6 628	1 600	.34	.00959 125	702	.84	6 841	809
.35	.00800 666	503	.85	8 228	602	.35	.00960 827	705	.85	.01048 650	811
.36	2 169	505	.86	.00879 830	604	.36	2 532	707	.86	.01050 461	813
.37	3 674	507	.87	.00881 434	607	.37	4 239	709	.87	2 274	815
.38	5 181	509	.88	3 041	608	.38	5 948	710	.88	4 089	818
.39	6 690	511	.89	4 649	610	.39	7 658	713	.89	5 907	819
.40	8 201	513	.90	6 259	612	.40	.00969 371	716	.90	7 726	822
.41	.00809 714	514	.91	7 871	614	.41	.00971 087	717	.91	.01059 548	824
.42	.00811 228	517	.92	.00889 485	616	.42	2 804	719	.92	.01061 372	826
.43	2 745	519	.93	.00891 101	618	.43	4 523	721	.93	3 198	829
.44	4 264	520	.94	2 719	620	.44	6 244	724	.94	5 027	830
.45	5 784	523	.95	4 339	623	.45	7 968	726	.95	6 857	833
.46	7 307	525	.96	5 962	624	.46	.00979 694	727	.96	.01068 690	835
.47	.00818 832	526	.97	7 586	626	.47	.00981 421	730	.97	.01070 525	837
.48	.00820 358	528	.98	.00899 212	629	.48	3 151	732	.98	2 362	840
.49	1 886	1 531	.99	.00900 841	1 630	.49	4 883	1 734	.99	4 202	1 841
.50	.00823 417		1.00	.00902 471		.50	.00986 617		1.00	.01076 043	

INVOLUTE FUNCTIONS

18° Φ	INV Φ	DIFF	18° Φ	INV Φ	DIFF	19° Φ	INV Φ	DIFF	19° Φ	INV Φ	DIFF
.00	.01076 043	1 844	.50	.01170 941	1 955	.00	.01271 506	2 070	.50	.01377 937	2 190
.01	7 887	846	.51	2 896	957	.01	3 576	073	.51	.01380 127	192
.02	.01079 733	848	.52	4 853	960	.02	5 649	075	.52	2 319	195
.03	.01081 581	850	.53	6 813	962	.03	7 724	078	.53	4 514	197
.04	3 431	852	.54	.01178 775	964	.04	.01279 802	079	.54	6 711	2 199
.05	5 283	855	.55	.01180 739	966	.05	.01281 881	083	.55	.01388 910	2 202
.06	7 138	857	.56	2 705	969	.06	3 964	084	.56	.01391 112	205
.07	.01088 995	859	.57	4 674	971	.07	6 048	087	.57	3 317	207
.08	.01090 854	861	.58	6 645	974	.08	8 135	090	.58	5 524	209
.09	2 715	864	.59	.01188 619	975	.09	.01290 225	091	.59	7 733	212
.10	4 579	865	.60	.01190 594	978	.10	.01292 316	094	.60	.01399 945	214
.11	6 444	868	.61	2 572	980	.11	4 410	096	.61	.01402 159	217
.12	.01098 312	870	.62	4 552	982	.12	6 506	2 099	.62	4 376	219
.13	.01100 182	873	.63	6 534	985	.13	8 605	2 102	.63	6 595	222
.14	2 055	874	.64	.01198 519	987	.14	.01300 707	103	.64	.01408 817	224
.15	3 929	877	.65	.01200 506	989	.15	.01302 810	106	.65	.01411 041	226
.16	5 806	879	.66	2 495	992	.16	4 916	108	.66	3 267	229
.17	7 685	881	.67	4 487	994	.17	7 024	110	.67	5 496	231
.18	.01109 566	883	.68	6 481	996	.18	.01309 134	113	.68	7 727	234
.19	.01111 449	886	.69	.01208 477	1 999	.19	.01311 247	116	.69	.01419 961	236
.20	3 335	888	.70	.01210 476	2 000	.20	3 363	118	.70	.01422 197	239
.21	5 223	889	.71	2 476	003	.21	5 481	120	.71	4 436	241
.22	7 112	893	.72	4 479	006	.22	7 601	122	.72	6 677	244
.23	.01119 005	894	.73	6 485	007	.23	.01319 723	125	.73	.01428 921	246
.24	.01120 899	897	.74	.01218 492	010	.24	.01321 848	127	.74	.01431 167	249
.25	2 796	1 899	.75	.01220 502	013	.25	3 975	130	.75	3 416	251
.26	4 695	1 901	.76	2 515	014	.26	6 105	132	.76	5 667	254
.27	6 596	903	.77	4 529	017	.27	.01328 237	134	.77	.01437 921	256
.28	.01128 499	906	.78	6 546	019	.28	.01330 371	137	.78	.01440 177	258
.29	.01130 405	908	.79	.01228 565	022	.29	2 508	139	.79	2 435	261
.30	2 313	910	.80	.01230 587	024	.30	4 647	142	.80	4 696	264
.31	4 223	912	.81	2 611	026	.31	6 789	144	.81	6 960	266
.32	6 135	715	.82	4 637	029	.32	.01338 933	146	.82	.01449 226	268
.33	8 050	916	.83	6 666	030	.33	.01341 079	149	.83	.01451 494	271
.34	.01139 966	919	.84	.01238 696	033	.34	3 228	151	.84	3 765	273
.35	.01141 885	922	.85	.01240 729	036	.35	5 379	154	.85	6 038	276
.36	3 807	923	.86	2 765	037	.36	7 533	156	.86	.01458 314	279
.37	5 730	926	.87	4 802	040	.37	.01349 689	158	.87	.01460 593	280
.38	7 656	928	.88	6 842	043	.38	.01351 847	161	.88	2 873	284
.39	.01147 584	930	.89	.01248 885	045	.39	4 008	164	.89	5 157	286
.40	.01151 514	933	.90	.01250 930	047	.40	6 172	165	.90	7 443	288
.41	3 447	935	.91	2 977	049	.41	.01358 337	168	.91	.01469 731	291
.42	5 382	936	.92	5 026	052	.42	.01360 505	171	.92	.01472 022	293
.43	7 318	940	.93	7 078	054	.43	2 676	173	.93	4 315	296
.44	.01159 258	941	.94	.01259 132	056	.44	4 849	175	.94	6 611	2 298
.45	.01161 199	944	.95	.01261 188	059	.45	7 024	178	.95	.01478 909	2 301
.46	3 143	946	.96	3 247	061	.46	.01369 202	180	.96	.01481 210	303
.47	5 089	948	.97	5 308	064	.47	.01371 382	182	.97	3 513	306
.48	7 037	951	.98	7 372	065	.48	3 564	185	.98	5 819	309
.49	.01168 988	1 953	.99	.01269 437	2 069	.49	5 749	2 188	.99	.01488 128	2 310
.50	.01170 941		1.00	.01271 506		.50	.01377 937		1.00	.01490 438	

INVOLUTE FUNCTIONS

20° Φ	INV Φ	DIFF.	20° Φ	INV Φ	DIFF.	21° Φ	INV Φ	DIFF.	21° Φ	INV Φ	DIFF.
.00	.01490 438	2 314	.50	.01609 218	2 441	.00	.01734 489	2 573	.50	.01866 469	2 709
.01	2 752	316	.51	11 659	444	.01	37 062	576	.51	69 178	713
.02	5 068	318	.52	14 103	446	.02	39 638	579	.52	71 891	715
.03	7 386	321	.53	16 549	449	.03	42 217	581	.53	74 606	717
.04	.01499 707	323	.54	18 998	452	.04	44 798	584	.54	77 323	721
.05	.01502 030	326	.55	21 450	454	.05	47 382	586	.55	80 044	724
.06	4 356	329	.56	23 904	457	.06	49 968	589	.56	82 768	726
.07	6 685	331	.57	26 361	459	.07	52 557	592	.57	85 494	729
.08	.01509 016	333	.58	28 820	462	.08	55 149	595	.58	88 223	732
.09	.01511 349	336	.59	31 282	464	.09	57 744	2 597	.59	90 955	734
.10	3 685	339	.60	33 746	467	.10	60 341	2 600	.60	93 689	737
.11	6 024	341	.61	36 213	470	.11	62 941	603	.61	96 426	741
.12	.01518 365	344	.62	38 683	473	.12	65 544	606	.62	.01899 167	743
.13	.01520 709	346	.63	41 156	475	.13	68 150	608	.63	.01901 910	745
.14	3 055	349	.64	43 631	477	.14	70 758	611	.64	04 655	749
.15	5 404	351	.65	46 108	480	.15	73 369	613	.65	07 404	751
.16	.01527 755	354	.66	48 588	483	.16	75 982	617	.66	10 155	754
.17	.01530 109	356	.67	51 071	486	.17	78 599	619	.67	12 909	757
.18	2 465	359	.68	53 557	488	.18	81 218	621	.68	15 666	760
.19	4 824	361	.69	56 045	491	.19	83 839	625	.69	18 426	762
.20	7 185	364	.70	58 536	493	.20	86 464	627	.70	21 188	766
.21	.01539 549	367	.71	61 029	496	.21	89 091	630	.71	23 954	768
.22	.01541 916	369	.72	63 525	2 499	.22	91 721	632	.72	26 722	771
.23	4 285	371	.73	66 024	2 501	.23	94 353	636	.73	29 493	774
.24	6 656	374	.74	68 525	504	.24	96 989	638	.74	32 267	776
.25	.01549 030	377	.75	71 029	506	.25	.01799 627	640	.75	35 043	780
.26	.01551 407	379	.76	73 535	510	.26	.01802 267	644	.76	37 823	782
.27	3 786	382	.77	76 045	512	.27	04 911	646	.77	40 605	785
.28	6 168	385	.78	78 557	514	.28	07 557	649	.78	43 390	788
.29	.01558 553	386	.79	81 071	517	.29	10 206	652	.79	46 178	791
.30	.01560 939	390	.80	83 588	520	.30	12 858	654	.80	48 969	793
.31	3 329	392	.81	86 108	522	.31	15 512	657	.81	51 762	797
.32	5 721	395	.82	88 630	525	.32	18 169	660	.82	54 559	2 799
.33	.01568 116	2 397	.83	91 155	528	.33	20 829	663	.83	57 358	2 802
.34	.01570 513	2 400	.84	93 683	531	.34	23 492	665	.84	60 160	805
.35	2 913	402	.85	96 214	533	.35	26 157	668	.85	62 965	807
.36	5 315	405	.86	.01698 747	535	.36	28 825	671	.86	65 772	811
.37	.01577 720	407	.87	.01701 282	539	.37	31 496	674	.87	68 583	813
.38	.01580 127	410	.88	03 821	541	.38	34 170	676	.88	71 396	817
.39	2 537	413	.89	06 362	543	.39	36 846	679	.89	74 213	819
.40	4 950	415	.90	08 905	547	.40	39 525	682	.90	77 032	822
.41	7 365	418	.91	11 452	549	.41	42 207	685	.91	79 854	824
.42	.01589 783	420	.92	14 001	551	.42	44 892	687	.92	82 678	828
.43	.01592 203	423	.93	16 552	555	.43	47 579	691	.93	85 506	830
.44	4 626	426	.94	19 107	557	.44	50 270	693	.94	88 336	834
.45	7 052	428	.95	21 664	560	.45	52 963	695	.95	91 170	836
.46	.01599 480	431	.96	24 224	562	.46	55 658	2 699	.96	94 006	839
.47	.01601 911	433	.97	26 786	565	.47	58 357	2 701	.97	96 845	842
.48	4 344	436	.98	29 351	568	.48	61 058	704	.98	.01999 687	844
.49	6 780	2 438	.99	31 919	2 570	.49	63 762	2 707	.99	.02002 531	2 848
.50	.01609 218		1.00	.01734 489		.50	.01866 469		1.00	.02005 379	

INVOLUTE FUNCTIONS

22°			22°			23°			23°		
φ	INV φ	DIFF.	φ	INV φ	DIFF.	φ	INV φ	DIFF.	φ	INV φ	DIFF.
.00	.02005 379	2 851	.50	.02151 448	2 996	.00	.02304 909	3 146	.50	.02466 000	3 301
.01	08 230	853	.51	54 444	2 999	.01	08 055	149	.51	69 301	305
.02	11 083	856	.52	57 443	3 002	.02	11 204	153	.52	72 606	308
.03	13 939	859	.53	60 445	005	.03	14 357	155	.53	75 914	310
.04	16 798	862	.54	63 450	008	.04	17 512	159	.54	79 224	314
.05	19 660	865	.55	66 458	010	.05	20 671	161	.55	82 538	317
.06	22 525	868	.56	69 468	014	.06	23 832	165	.56	85 855	321
.07	25 393	870	.57	72 482	017	.07	26 997	167	.57	89 176	323
.08	28 263	873	.58	75 499	020	.08	30 164	171	.58	92 499	327
.09	31 136	877	.59	78 519	022	.09	33 335	174	.59	95 826	329
.10	34 013	879	.60	81 541	026	.10	36 509	177	.60	.02499 155	333
.11	36 892	882	.61	84 567	029	.11	39 686	180	.61	.02502 488	336
.12	39 774	885	.62	87 596	031	.12	42 866	183	.62	05 824	340
.13	42 659	888	.63	90 627	035	.13	46 049	186	.63	09 164	342
.14	45 547	891	.64	93 662	037	.14	49 235	189	.64	12 506	346
.15	48 438	893	.65	96 699	041	.15	52 424	192	.65	15 852	349
.16	51 331	897	.66	.02199 740	044	.16	55 616	196	.66	19 201	352
.17	54 228	2 899	.67	.02202 784	046	.17	58 812	198	.67	22 553	355
.18	57 127	2 903	.68	05 830	050	.18	62 010	202	.68	25 908	358
.19	60 030	905	.69	08 880	052	.19	65 212	204	.69	29 266	362
.20	62 935	908	.70	11 932	056	.20	68 416	208	.70	32 628	364
.21	65 843	911	.71	14 988	058	.21	71 624	211	.71	35 992	368
.22	68 754	914	.72	18 046	062	.22	74 835	213	.72	39 360	372
.23	71 668	917	.73	21 108	064	.23	78 048	217	.73	42 732	374
.24	74 585	919	.74	24 172	068	.24	81 265	220	.74	46 106	377
.25	77 504	923	.75	27 240	070	.25	84 485	224	.75	49 483	381
.26	80 427	926	.76	30 310	074	.26	87 709	226	.76	52 864	384
.27	83 353	928	.77	33 384	076	.27	90 935	229	.77	56 248	387
.28	86 281	931	.78	36 460	080	.28	94 164	233	.78	59 635	390
.29	89 212	935	.79	39 540	082	.29	.02397 397	235	.79	63 025	394
.30	92 147	937	.80	42 622	086	.30	.02400 632	239	.80	66 419	3 397
.31	95 084	940	.81	45 708	089	.31	03 871	242	.81	69 816	3 400
.32	.02098 024	943	.82	48 797	091	.32	07 113	245	.82	73 216	403
.33	.02100 967	946	.83	51 888	095	.33	10 358	248	.83	76 619	406
.34	03 913	949	.84	54 983	3 097	.34	13 606	251	.84	80 025	410
.35	06 862	952	.85	58 080	3 101	.35	16 857	254	.85	83 435	413
.36	09 814	955	.86	61 181	104	.36	20 111	258	.86	86 848	416
.37	12 769	957	.87	64 285	106	.37	23 367	260	.87	90 264	419
.38	15 726	961	.88	67 391	110	.38	26 629	264	.88	93 683	423
.39	18 687	963	.89	70 501	113	.39	29 893	267	.89	.02597 106	425
.40	21 650	967	.90	73 614	116	.40	33 160	270	.90	.02600 531	429
.41	24 617	969	.91	76 730	119	.41	36 430	273	.91	03 960	432
.42	27 586	973	.92	79 849	122	.42	39 703	276	.92	07 392	436
.43	30 559	975	.93	82 971	124	.43	42 979	279	.93	10 828	439
.44	33 534	978	.94	86 095	128	.44	46 268	283	.94	14 267	441
.45	36 512	982	.95	89 223	131	.45	49 541	285	.95	17 708	446
.46	39 494	984	.96	92 354	134	.46	52 826	289	.96	21 154	448
.47	42 478	987	.97	95 488	138	.47	56 115	292	.97	24 602	452
.48	45 465	990	.98	.02298 626	140	.48	59 407	295	.98	28 054	454
.49	48 455	2 993	.99	.02301 766	3 143	.49	62 702	3 298	.99	31 508	3 458
.50	.02151 448		1.00	.02304 909		.50	.02466 000		1.00	.02634 966	

INVOLUTE FUNCTIONS

24° Φ	INV Φ	DIFF	24° Φ	INV Φ	DIFF	25° Φ	INV Φ	DIFF	25° Φ	INV Φ	DIFF
.00	.02634 966	3 462	.50	.02812 059	3 626	.00	.02997 535	3 796	.50	.03191 657	3 973
.01	38 428	464	.51	15 685	630	.01	.03001 331	3 801	.51	95 630	976
.02	41 892	468	.52	19 315	633	.02	05 132	803	.52	.03199 606	980
.03	45 360	472	.53	22 948	637	.03	08 935	808	.53	.03203 586	983
.04	48 832	474	.54	26 585	640	.04	12 743	810	.54	07 569	987
.05	52 306	478	.55	30 225	643	.05	16 553	814	.55	11 556	990
.06	55 784	480	.56	33 868	647	.06	20 367	818	.56	15 546	994
.07	59 264	485	.57	37 515	650	.07	24 185	821	.57	19 540	3 997
.08	62 749	487	.58	41 165	653	.08	28 006	825	.58	23 537	4 002
.09	66 236	491	.59	44 818	657	.09	31 831	828	.59	27 539	004
.10	69 727	494	.60	48 475	660	.10	35 659	831	.60	31 543	009
.11	73 221	3 497	.61	52 135	664	.11	39 490	835	.61	35 552	012
.12	76 718	3 501	.62	55 799	667	.12	43 325	839	.62	39 564	015
.13	80 219	503	.63	59 466	670	.13	47 164	842	.63	43 579	019
.14	83 722	507	.64	63 136	673	.14	51 006	845	.64	47 598	023
.15	87 229	511	.65	66 809	678	.15	54 851	849	.65	51 621	026
.16	90 740	513	.66	70 487	680	.16	58 700	853	.66	55 647	030
.17	94 253	517	.67	74 167	684	.17	62 553	856	.67	59 677	033
.18	.02697 770	521	.68	77 851	687	.18	66 409	859	.68	63 710	037
.19	.02701 291	523	.69	81 538	691	.19	70 268	863	.69	67 747	041
.20	04 814	527	.70	85 229	694	.20	74 131	866	.70	71 788	044
.21	08 341	530	.71	88 923	3 697	.21	77 997	870	.71	75 832	048
.22	11 871	533	.72	92 620	3 701	.22	81 867	874	.72	79 880	052
.23	15 404	537	.73	.02896 321	704	.23	85 741	877	.73	83 932	055
.24	18 941	540	.74	.02900 025	707	.24	89 618	880	.74	87 987	059
.25	22 481	543	.75	03 732	711	.25	93 498	884	.75	92 046	062
.26	26 024	547	.76	07 443	715	.26	.03097 382	888	.76	.03296 108	066
.27	29 571	550	.77	11 158	718	.27	.03101 270	891	.77	.03300 174	070
.28	33 121	553	.78	14 876	721	.28	05 161	894	.78	04 244	073
.29	36 674	556	.79	18 597	725	.29	09 055	3 898	.79	08 317	077
.30	40 230	560	.80	22 322	728	.30	12 953	3 902	.80	12 394	081
.31	43 790	563	.81	26 050	731	.31	16 855	905	.81	16 475	084
.32	47 353	567	.82	29 781	735	.32	20 760	909	.82	20 559	088
.33	50 920	570	.83	33 516	738	.33	24 669	912	.83	24 647	091
.34	54 490	573	.84	37 254	742	.34	28 581	916	.84	28 738	095
.35	58 063	576	.85	40 996	745	.35	32 497	919	.85	32 833	4 099
.36	61 639	580	.86	44 741	749	.36	36 416	923	.86	36 932	4 102
.37	65 219	583	.87	48 490	752	.37	40 339	926	.87	41 034	106
.38	68 802	586	.88	52 242	755	.38	44 265	930	.88	45 140	110
.39	72 388	590	.89	55 997	759	.39	48 195	933	.89	49 250	113
.40	75 978	593	.90	59 756	763	.40	52 128	937	.90	53 363	117
.41	79 571	597	.91	63 519	765	.41	56 065	941	.91	57 480	121
.42	83 168	3 599	.92	67 284	770	.42	60 006	944	.92	61 601	124
.43	86 767	3 603	.93	71 054	772	.43	63 950	947	.93	65 725	128
.44	90 370	607	.94	74 826	776	.44	67 897	951	.94	69 853	132
.45	93 977	609	.95	78 602	780	.45	71 848	955	.95	73 985	135
.46	.02797 586	614	.96	82 382	783	.46	75 803	958	.96	78 120	139
.47	.02801 200	616	.97	86 165	786	.47	79 761	962	.97	82 259	143
.48	04 816	620	.98	89 951	790	.48	83 723	965	.98	86 402	146
.49	08 436	3 623	.99	93 741	3 794	.49	87 688	3 969	.99	90 548	4 150
.50	.02812 059		1.00	.02997 535		.50	.03191 657		1.00	.03394 698	

INVOLUTE FUNCTIONS

26°			26°			27°			27°		
φ	INV φ	DIFF	φ	INV φ	DIFF	φ	INV φ	DIFF	φ	INV φ	DIFF
.00	.03394 698	4 154	.50	.03606 936	4 340	.00	.03828 655	4 533	.50	.04060 151	4 731
.01	.03398 852	157	.51	11 276	345	.01	33 188	537	.51	64 882	736
.02	.03403 009	161	.52	15 621	348	.02	37 725	541	.52	69 618	740
.03	07 170	165	.53	19 969	351	.03	42 266	545	.53	74 358	744
.04	11 335	169	.54	24 320	356	.04	46 811	549	.54	79 102	747
.05	15 504	172	.55	28 676	360	.05	51 360	553	.55	83 849	752
.06	19 676	175	.56	33 036	363	.06	55 913	556	.56	88 601	756
.07	23 851	180	.57	37 399	367	.07	60 469	561	.57	93 357	760
.08	28 031	183	.58	41 766	371	.08	65 030	564	.58	.04098 117	764
.09	32 214	187	.59	46 137	375	.09	69 594	569	.59	.04102 881	768
.10	36 401	191	.60	50 512	378	.10	74 163	572	.60	07 649	772
.11	40 592	194	.61	54 890	383	.11	78 735	576	.61	12 421	777
.12	44 786	4 198	.62	59 273	386	.12	83 311	580	.62	17 198	780
.13	48 984	4 202	.63	63 659	390	.13	87 891	585	.63	21 978	784
.14	53 186	205	.64	68 049	394	.14	92 476	588	.64	26 762	789
.15	57 391	209	.65	72 443	4 397	.15	.03897 064	592	.65	31 551	792
.16	61 600	213	.66	76 840	4 402	.16	03901 656	4 596	.66	36 343	4 797
.17	65 813	217	.67	81 242	405	.17	06 252	4 600	.67	41 140	4 800
.18	70 030	220	.68	85 647	409	.18	10 852	604	.68	45 940	805
.19	74 250	224	.69	90 056	413	.19	15 456	608	.69	50 745	808
.20	78 474	228	.70	94 469	417	.20	20 064	611	.70	55 553	813
.21	82 702	231	.71	.03698 886	421	.21	24 675	616	.71	60 366	817
.22	86 933	235	.72	.03703 307	424	.22	29 291	620	.72	65 183	821
.23	91 168	239	.73	07 731	429	.23	33 911	624	.73	70 004	825
.24	95 407	243	.74	12 160	432	.24	38 535	627	.74	74 829	829
.25	.03499 650	246	.75	16 592	436	.25	43 162	632	.75	79 658	834
.26	.03503 896	250	.76	21 028	440	.26	47 794	636	.76	84 492	837
.27	08 146	254	.77	25 468	444	.27	52 430	639	.77	89 329	842
.28	12 400	258	.78	29 912	447	.28	57 069	644	.78	94 171	845
.29	16 658	261	.79	34 359	452	.29	61 713	647	.79	.04199 016	850
.30	20 919	265	.80	38 811	455	.30	66 360	652	.80	.04203 866	853
.31	25 184	269	.81	43 266	459	.31	71 012	655	.81	08 719	858
.32	29 453	273	.82	47 725	464	.32	75 667	660	.82	13 577	862
.33	33 726	276	.83	52 189	467	.33	80 327	663	.83	18 439	866
.34	38 002	280	.84	56 656	470	.34	84 990	668	.84	23 305	871
.35	42 282	284	.85	61 126	475	.35	89 658	671	.85	28 176	874
.36	46 566	287	.86	65 601	479	.36	94 329	676	.86	33 050	878
.37	50 853	292	.87	70 080	482	.37	.03999 005	679	.87	37 928	883
.38	55 145	295	.88	74 562	486	.38	.04003 684	684	.88	42 811	887
.39	59 440	4 299	.89	79 048	491	.39	08 368	687	.89	47 698	890
.40	63 739	4 303	.90	83 539	494	.40	13 055	692	.90	52 588	895
.41	68 042	306	.91	88 033	4 498	.41	17 747	4 695	.91	57 483	4 899
.42	72 348	310	.92	92 531	4 502	.42	22 442	4 700	.92	62 382	4 904
.43	76 658	314	.93	.03797 033	506	.43	27 142	703	.93	67 286	907
.44	80 972	318	.94	.03801 539	509	.44	31 845	708	.94	72 193	911
.45	85 290	322	.95	06 048	514	.45	36 553	711	.95	77 104	916
.46	89 612	325	.96	10 562	517	.46	41 264	716	.96	82 020	920
.47	93 937	329	.97	15 079	522	.47	45 980	719	.97	86 940	924
.48	.03598 266	333	.98	19 601	525	.48	50 699	723	.98	91 864	928
.49	03602 599	4 337	.99	24 126	4 529	.49	55 422	4 729	.99	.04296 792	4 932
.50	03606 936		100	.03828 655		.50	.04060 151		1.00	.04301 724	

INVOLUTE FUNCTIONS

28° φ	INV φ	DIFF.	28° φ	INV φ	DIFF.	29° φ	INV φ	DIFF.	29° φ	INV φ	DIFF.
.00	.04301 724	4 937	.50	.04553 686	5 148	.00	.04816 357	5 365	.50	.05090 065	5 589
.01	06 661	940	.51	58 834	151	.01	21 722	369	.51	.05095 654	594
.02	11 601	945	.52	63 985	156	.02	27 091	374	.52	.05101 248	5 598
.03	16 546	949	.53	69 141	161	.03	32 465	378	.53	06 846	5 602
.04	21 495	953	.54	74 302	164	.04	37 843	382	.54	12 448	608
.05	26 448	957	.55	79 466	169	.05	43 225	387	.55	18 056	612
.06	31 405	961	.56	84 635	173	.06	48 612	392	.56	23 668	616
.07	36 366	966	.57	89 808	178	.07	54 004	5 396	.57	29 284	621
.08	41 332	969	.58	.04594 986	181	.08	59 400	5 400	.58	34 905	626
.09	46 301	974	.59	.04600 167	186	.09	64 800	405	.59	40 531	630
.10	51 275	978	.60	05 353	191	.10	70 205	409	.60	46 161	634
.11	56 253	983	.61	10 544	194	.11	75 614	414	.61	51 795	640
.12	61 236	986	.62	15 738	5 199	.12	81 028	418	.62	57 435	644
.13	66 222	991	.63	20 937	5 204	.13	86 446	422	.63	63 079	648
.14	71 213	995	.64	26 141	207	.14	91 868	427	.64	68 727	653
.15	76 208	4 999	.65	31 348	212	.15	.04897 295	432	.65	74 380	658
.16	81 207	5 003	.66	36 560	217	.16	.04902 727	435	.66	80 038	662
.17	86 210	007	.67	41 777	220	.17	08 162	441	.67	85 700	667
.18	91 217	012	.68	46 997	225	.18	13 603	445	.68	91 367	671
.19	.04396 229	016	.69	52 222	229	.19	19 048	449	.69	.05197 038	676
.20	.04401 245	020	.70	57 451	234	.20	24 497	454	.70	.05202 714	681
.21	06 265	024	.71	62 685	238	.21	29 951	458	.71	08 395	685
.22	11 289	028	.72	67 923	242	.22	35 409	462	.72	14 080	690
.23	16 317	033	.73	73 165	247	.23	40 871	468	.73	19 770	694
.24	21 350	037	.74	78 412	251	.24	46 339	471	.74	25 464	5 699
.25	26 387	041	.75	83 663	255	.25	51 810	476	.75	31 163	5 704
.26	31 428	045	.76	88 918	260	.26	57 286	481	.76	36 867	709
.27	36 473	050	.77	94 178	264	.27	62 767	485	.77	42 576	713
.28	41 523	054	.78	.04699 442	268	.28	68 252	490	.78	48 289	717
.29	46 577	058	.79	.04704 710	273	.29	73 742	494	.79	54 006	722
.30	51 635	062	.80	09 983	277	.30	79 236	5 498	.80	59 728	727
.31	56 697	066	.81	15 260	281	.31	84 734	5 503	.81	65 455	732
.32	61 763	071	.82	20 541	286	.32	90 237	508	.82	71 187	736
.33	66 834	075	.83	25 827	290	.33	.04995 745	512	.83	76 923	741
.34	71 909	079	.84	31 117	295	.34	.05001 257	517	.84	82 664	745
.35	76 988	083	.85	36 412	5 299	.35	06 774	521	.85	88 409	750
.36	82 071	088	.86	41 711	5 303	.36	12 295	525	.86	94 159	755
.37	87 159	092	.87	47 014	308	.37	17 820	530	.87	.05299 914	759
.38	92 251	5 096	.88	52 322	312	.38	23 350	535	.88	.05305 673	764
.39	.04497 347	5 101	.89	57 634	316	.39	28 885	539	.89	11 437	769
.40	.04502 448	104	.90	62 950	321	.40	34 424	544	.90	17 206	773
.41	07 552	109	.91	68 271	326	.41	39 968	548	.91	22 979	778
.42	12 661	113	.92	73 597	329	.42	45 516	553	.92	28 757	783
.43	17 774	118	.93	78 926	334	.43	51 069	557	.93	34 540	787
.44	22 892	122	.94	84 260	339	.44	56 626	562	.94	40 327	792
.45	28 014	126	.95	89 599	342	.45	62 188	566	.95	46 119	5 797
.46	33 140	130	.96	.04794 941	348	.46	67 754	571	.96	51 916	5 801
.47	38 270	134	.97	.04800 289	351	.47	73 325	575	.97	57 717	806
.48	43 404	139	.98	05 640	356	.48	78 900	580	.98	63 523	811
.49	48 543	5 143	.99	10 996	5 361	.49	84 480	5 585	.99	69 334	5 815
.50	.04553 686		1.00	.04816 357		.50	.05090 065		1.00	.05375 149	

INVOLUTE FUNCTIONS

30°			30°			31°			31°		
Φ	INV Φ	DIFF.	Φ	INV Φ	DIFF.	Φ	INV Φ	DIFF.	Φ	INV Φ	DIFF
.00	.05375 149	5 821	.50	.05671 959	6 059	.00	.05980 855	6 304	.50	.06302 207	6 557
.01	80 970	824	.51	78 018	063	.01	87 159	309	51	08 764	562
.02	86 794	830	.52	84 081	068	.02	93 468	313	.52	15 326	567
.03	92 624	834	.53	90 149	073	.03	.05999 781	319	.53	21 893	572
.04	.05398 458	839	.54	.05696 222	077	.04	.06006 100	324	.54	28 465	577
.05	.05404 297	844	.55	.05702 299	083	.05	12 424	328	.55	35 042	583
.06	10 141	848	.56	08 382	087	.06	18 752	334	56	41 625	587
.07	15 989	853	.57	14 469	092	.07	25 086	339	.57	48 212	593
.08	21 842	858	.58	20 561	6 097	.08	31 425	343	.58	54 805	6 598
.09	27 700	862	.59	26 658	6 102	.09	37 768	349	.59	61 403	6 603
.10	33 562	867	.60	32 760	107	.10	44 117	354	.60	68 006	608
.11	39 429	872	.61	38 867	111	.11	50 471	358	.61	74 614	614
.12	45 301	877	.62	44 978	117	.12	56 829	364	.62	81 228	618
.13	51 178	881	.63	51 095	121	.13	63 193	369	.63	87 846	624
.14	57 059	886	.64	57 216	127	.14	69 562	374	.64	.06394 470	629
.15	62 945	891	.65	63 343	131	.15	75 936	379	.65	.06401 099	634
.16	68 836	5 895	.66	69 474	136	.16	82 315	383	.66	07 733	639
.17	74 731	5 901	.67	75 610	141	.17	88 698	389	.67	14 372	645
.18	80 632	905	.68	81 751	145	.18	.06095 087	394	.68	21 017	650
19	86 537	909	.69	87 896	151	.19	.06101 481	6 399	.69	27 667	655
.20	92 446	915	.70	.05794 047	156	.20	07 880	6 404	.70	34 322	660
.21	.05498 361	919	.71	.05800 203	160	.21	.06114 284	409	.71	40 982	665
.22	.05504 280	924	.72	06 363	165	.22	20 693	414	.72	47 647	670
.23	10 204	929	.73	12 528	171	.23	27 107	419	.73	54 317	676
.24	16 133	933	.74	18 699	175	.24	33 526	425	.74	60 993	681
.25	22 066	939	.75	24 874	180	.25	39 951	429	.75	67 674	686
.26	28 005	943	.76	31 054	185	.26	46 380	434	.76	74 360	691
.27	33 948	948	.77	37 239	190	.27	52 814	439	.77	81 051	6 697
.28	39 896	952	.78	43 429	6 194	.28	59 253	445	.78	87 748	6 702
.29	45 848	958	.79	49 623	6 200	.29	65 698	449	.79	.06494 450	707
.30	51 806	962	.80	55 823	205	.30	72 147	455	.80	.06501 157	712
.31	57 768	967	.81	62 028	209	.31	78 602	460	.81	07 869	717
.32	63 735	971	.82	68 237	215	.32	85 062	464	.82	14 586	723
.33	69 706	977	.83	74 452	219	.33	91 526	470	.83	21 309	728
.34	75 683	981	.84	80 671	225	.34	.06197 996	475	.84	28 037	733
.35	81 664	986	.85	86 896	229	.35	.06204 471	480	.85	34 770	739
.36	87 650	991	.86	93 125	234	.36	10 951	485	.86	41 509	743
.37	93 641	5 996	.87	.05899 359	239	.37	17 436	491	.87	48 252	749
.38	.05599 637	6 000	.88	.05905 598	244	.38	23 927	6 495	.88	55 001	754
.39	.05605 637	005	.89	11 842	249	.39	30 422	6 500	.89	61 755	760
.40	11 642	010	.90	18 091	255	.40	36 922	506	90	68 515	765
.41	17 652	015	.91	24 346	259	.41	43 428	510	.91	75 280	770
.42	23 667	020	.92	30 605	263	.42	49 938	516	.92	82 050	775
.43	29 687	024	.93	36 868	269	.43	56 454	521	.73	88 825	780
.44	35 711	030	.94	43 137	274	.44	62 975	526	.94	.06545 605	786
.45	41 741	034	.95	49 411	279	.45	69 501	531	95	.06602 391	791
.46	47 775	039	.96	55 690	284	.46	76 032	536	.96	09 182	6 796
.47	53 814	043	.97	61 974	289	.47	82 568	541	.97	15 978	6 802
.48	59 857	049	.98	68 263	293	.48	89 109	547	.98	22 780	807
.49	65 906	6 053	.99	74 556	6 299	.49	.06295 656	6 551	.99	29 587	6 812
.50	.05671 959		1.00	.05980 855		50	.06302 207		1.00	06636 399	

Reproduced with permission from Buckingham's Manual of Gear Design, Vol. III, Section One, by Earle Buckingham, revised by Eliot K. Buckingham, published by Buckingham Assoc, Inc., Springfield, VT, 1981).

References

AGMA Standard 250.04 (Sept., 1981), "Lubrication of Industrial Enclosed Gear Drives."

AGMA Standard 251.02 (Nov., 1974), "Lubrication of Industrial Open Gearing."

AGMA 341.02-1965 (R1970), "Design of General Industrial Course-Pitch Cylindrical Wormgearing."

AGMA Standard 390.03 (Jan., 1971), "Gear Handbook, Vol. 1, Gear Classification, Materials and Measuring Methods for Unassembled Gears."

AGMA Standard 390.03A, (June, 1988), "Gear Handbook, Gear Classification, Materials & Measuring Methods for Bevel, Hypoid, Fine Pitch Worm Gearing and Racks, Only as Unassembled Gears."

ANSI/AGMA 100.04-1980 (R1989), "Nomenclature of Gear Tooth Failure Modes."

ANSI/AGMA 201.02-1968 (R1974), "Tooth Proportions for Coarse-Pitch Involute Spur Gears."

ANSI/AGMA 1012-F90, "Gear Nomenclature, Definitions of Terms with Symbols."

ANSI/AGMA 2001-B88 (Revision of 218.01 1982), "Fundamental Rating Factors and Calculation Methods for Involute Spur and Helical Gear Teeth."

ANSI/AGMA 2000-A88 (Mar., 1988), "Gear Classification and Inspection Handbook, Tolerances and Measuring Methods for Unassembled Spur and Helical Gears."

ANSI/AGMA 2002-B88 (Revision of 231.52), "Tooth Thickness Specification and Measurement."

ANSI/AGMA 2004-B89 (Jan., 1989), "Gear Materials and Heat Treatment Manual."

Ashcroft, Geoffrey, Pfauter-Maag Cutting Tools, Loves Park, IL, "Disposable Cutting Tools for Gear Manufacturing," a paper presented at AGMA Manufacturing Symposium, Apr. 7–9, 1991, Chicago.

Ashcroft, Geoffrey and Cluff, Brian W., American Pfauter, Loves Park, IL, "High Efficiency Gear Hobbing," a paper presented at AGMA fall meeting, Oct. 23–25, 1991, Detroit.

Buckingham, Earle, *Analytical Mechanics of Gears*, McGraw-Hill, New York, 1949.

Buckingham, Earle, "Manual of Gear Design" (Revised), Section 2, revision by Eliot K. Buckingham, Buckingham and Associates, Springfield, VT., 1981

Buckingham, Earle, "Manual of Gear Design" (Revised), Section 3, Revision by Eliot K. Buckingham, Buckingham & Associates, Springfield, VT., 1981

Campbell, Harry L., *The Working, Heat Treating and Welding of Steel*, John Wiley & Sons, New York.

Dudley, Darle W., Editor, *Gear Handbook*, McGraw-Hill, New York, 1962.

Dudley, Darle W., *The Evolution of the Gear Art*, American Gear Manufacturers Association, 1969.

Dudley, Darle W., *Handbook of Practical Gear Design*, McGraw-Hill, New York, 1984.

Dudley, Darle W., "Gear Technology—Past, Present and Future," a paper presented to the International Conference on Gearing, Zhengzhow, Henan, China, 1988.

Ewert, Richard H., "Gearing—Basic Theory and its Application—a Primer," Sewall Gear Mfg. Co., St. Paul, MN, 1980.

Ewert, Richard H., "Costs, their Determination, Allocation and Control," a consolidation of several papers presented at American Gear Manufacturers Association meetings, 1963 through 1985. A complete copy of this paper can be obtained by contacting the author at 1064 Lombard Ave., St. Paul, MN 55105-3255.

Ewert, Richard H., *Axial Vibration of Steam Turbine Buckets*, a thesis for partial requirements for BS and MS Degrees in Mechanical Engineering at MIT, 1938.

Lynwander, Peter, *Gear Drive Systems—Design and Application*, Marcel Dekker, New York, 1983.

Marks' Standard Handbook for Mechanical Engineers, 9th Edition, McGraw-Hill, New York, 1987, p. 8–36, Fig. 8.2.34, Tables 8.2.34 and 8.2.35.

Smith, J. Derek, *Gears and Their Vibration*, Marcel Dekker, Inc., New York, 1983.

Society of Automotive Engineers, Standard ANSI B92.1-1970, 7th Printing, 1988, "Involute Splines and Inspection."

South, David W. and Mancuso, Jon, Editors, *Mechanical Power Transmission Components*, Marcel Dekker, Inc., New York, 1994. (Chapter 7, "Gear Drives", by Richard H. Ewert.)

South, David W. and Ewert, Richard H., *Encyclopedic Dictionary of Gears and Gearing*, McGraw-Hill, New York, 1994.

Index